U0156784

生态沉湖

候鸟福地

The Ecology of Chenhu Lake
An Ideal Habitat for Migratory Birds

武汉市园林和林业局 组编

HAN BOOK 版 武汉出版社
WUHAN PUBLISHING HOUSE

编委会名单

前　言

武汉位于中国腹地、长江与汉水交汇处，水陆交通发达，具有连接东西、沟通南北的地理优势，素有“九省通衢”之称。“得中独厚”的地理区位、悠久的历史文化、湖泊密布的资源条件成为武汉的独特优势，早在20世纪之初就奠定了特大城市的空间框架。

两江交汇、三镇鼎立、百湖镶嵌，缘水而发、因水而兴、依水而美。市内现存湖泊166个，水域面积约2200 km^2，约占全市面积四分之一，湿地率和水体数量均居全国各大城市首位。“襟江、带河、链湖”的独特生态格局下形成了“人水和谐、水城共生”的“武汉模式”。

沉湖是武汉众多湖泊湿地中唯一的国际重要湿地。作为古云梦泽的遗存，在千百年的历史长卷中，其完整的生态系统、季节性的水文过程、优美的自然景观、丰富的生物多样性，堪为长江中下游湖泊群中的典型代表，也是东亚—澳大利西亚迁飞区南来北往候鸟的福地。

在《湿地公约》第十四届缔约方大会召开之际，作为会议参观考察点，相信沉湖这颗江汉明珠，将向全世界展现一幅人与自然和谐共生的“中国绿”画卷。

国际湿地公约科技委员会主席

北京林业大学教授

Preface

Wuhan is located in central China, where the Yangtze River meets its largest tributary, the Han River. As the "thoroughfare of nine provinces", with its land and water transportation networks well developed, the city serves as a hub connecting the east with the west, the north with the south. In the early 20th century, due to its unique location, long history, ancient culture, and densely distributed lakes, the spatial framework of a megacity was formed.

Comprised of three towns, separated by two converging rivers and dotted with so many lakes, the city has thrived, beautified by its water environments. It has 166 lakes, which occupy an area of about 2,200 square kilometers, about one-quarter of its total area, making Wuhan a major Chinese city with the highest percentage of wetland area and largest number of water bodies in China. These features also make Wuhan extremely unusual from a global perspective. Its unique ecological pattern of "being intertwined with rivers and lakes" has developed a "Wuhan model" defined by "harmony of humanity and water, co-existence of water and city".

Among the numerous lakes and wetlands in Wuhan, the Chenhu Lake Wetland is the only one of international importance. This wetland, a remnant of the ancient Yunmeng Wetland, is considered the representative of the lake groups in the Yangtze's middle and lower reaches in history thanks to its complete ecosystem, seasonal hydrological process, beautiful natural landscape and rich biodiversity. It also is an ideal habitat for migratory birds that traverse the East Asia–Australasia Flyway.

On the occasion of the 14th Meeting of the Conference of the Contracting Parties to the *Ramsar Convention on Wetlands* (COP14), it is believed that Chenhu Lake, the pearl of the Jianghan Plain, as a visiting site for the meeting, will present a picture of "China Green" with "harmonious coexistence of humanity and nature" to the world

Chair of the Scientific and Technical Review Panel of the Convention on Wetlands,
Professor of Beijing Forestry University
Dr. Lei Guangchun

目　录 Contents

第一章

Chapter I

江汉明珠

The Pearl of the Jianghan Plain

湿地是人类文明的摇篮。几千年前，我们的祖先“逐水草而居”，在与湿地相互依存的漫长历史过程中，创造了中华文明和湿地文化。河湖文化、稻作文化、渔猎文化是湿地文化的典型代表。在湿地与人类渔樵耕读的交汇演替中，承载和记录了人类的智慧、湿地的变迁和社会的演变。文明的演化告诉我们，生态兴则文明兴，人与自然是休戚与共的命运共同体。

Wetlands are a cradle of human civilization. Thousands of years ago, the ancestors of the Chinese lived in areas where water and grass were available, and during their long interdependence with wetlands, they managed to create the Chinese civilization boasting several “wetland cultures”, typified by river and lake culture, rice cultivation culture, and fishing and hunting culture. Through their integration with human food production and culture, wetlands bear the record of human knowledge and social evolution alongside their own changes. The history of civilization tells us that the prosperity of ecology means the prosperity of civilization, and that humanity and nature form a community with a shared future.

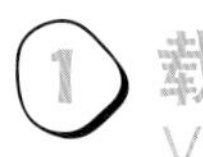

载沉载浮

Vicissitudes of Time

云梦泽畔孕育出的独特的楚文化，在大约2700年前形成。楚文化与北方文化表现出的不同特色，是大江大湖所赋予的鲜明个性。

沉湖湿地中的小道　李梓固摄
A Trail in the Chenhu Lake Wetland　Photo by Li Zigu

The Yunmeng Wetland nurtured the unique Chu Culture, which took form about 2,700 years ago. The distinctive character of this culture, different from that of Northern China, is shaped by the large rivers and lakes all around.

沉湖夕阳　李梓固摄
The Sunset over Chenhu Lake　Photo by Li Zigu

云梦泽最早见于《尚书•禹贡》：“云土梦作乂。”孔安国注曰：“云梦之泽在江南。”《尔雅•释地》曰：“鲁有大野，晋有大陆，秦有杨陓，宋有孟诸，楚有云梦。”郭璞注曰：“今南郡华容县东南巴丘湖是也。”华容古城在今湖北潜江西南。汉代文学家司马相如作的《子虚赋》中，对云梦泽作了极为宏丽的描绘。古云梦泽宽广，“方九百里”。

The expression “Yunmeng Wetland” was first mentioned in *The Book of Historical Documents*, and Kong Anguo noted that “the Yunmeng Wetland is to the south of the Yangtze River.” *Erya* ,the first Chinese Dictionary says, “The State of Lu has Daye, the State of Jin has Dalu, the State of Qin has Yangyu, the State of Song has Mengzhu, and the State of Chu has Yunmeng...” Guo Pu explained, “There is located present-day Baqiu Lake in the southeast of Huarong County, Nan Commandery.” Huarong Ancient

扫码观赏视频
Scan and watch

“云梦泽” 陈勇摄
The “Yunmeng Wetland” Photo by Chen Yong

City can be found southwest of Qianjiang, Hubei Province. Sima Xiangru, a great writer of the Han Dynasty, described the Yunmeng Wetland in his *Essay on Zi Xu* as "covering an area of about 81,000 square *li* (ancient Chinese miles)".

据史学家谭其骧先生所述，先秦到两汉、三国，云梦泽大致在江陵之东，江汉之间，华容县以南和以东。

According to Tan Qixiang, a historian in China, the Yunmeng Wetland, during the pre-Qin Dynasty, the Han Dynasty and the Three Kingdoms Period, was located roughly to the east of Jiangling County, south and east of Huarong County, between the Han River and the Yangtze.

冲积平原中的湖泽，往往变化非常频繁。长江和汉水的含沙量巨大，随着上游的逐步开发，江水挟带而来的沉积物与日俱增，水体逐渐缩小，陆地逐渐扩大。但同时江汉地区由于在地质构造运动中不断下降，水体缩小的趋势不大，甚至局部还会出现相反的情况。随着江汉间平原的日益扩展，到魏晋时期，云梦泽日益淤积、东移，其东端一直伸展到了大江东岸的沌阳县境。东晋或南朝初，华容东南的云梦泽主体由于大面积淤积，被分割成大浐、马骨、太白等湖沼陂池，"云梦泽"开始湮没于历史长河。南朝时，大浐、马骨二湖为最大，"夏水来则渺漭若海"。太白湖位于今蔡甸区南，约相当于今沉湖，《水经注》中的《江水注》《沔水注》皆有记载。

Lakes in alluvial plains tend to change very frequently in form. As the upper reaches of the Han River and the Yangtze gradually evolved, huge amounts of sand were released, the sediment in their waters increasing day by day, gradually shrinking the water bodies and expanding the surrounding lands. However, in the Jianghan area, geotectonic movement tended to lower the earth's surface, restricting and even reversing the shrinkage of the water bodies. With the expansion of the Jianghan Plain, sedimentation in the Yunmeng Wetland led to its gradual movement eastwards during the Wei and Jin dynasties, with its eastmost end stretching all the way into Zhuanyang County on the Yangtze's east bank. In the Eastern Jin Dynasty and early Southern Dynasties, the main body of the Yunmeng Wetland southeast of Huarong County was divided by silt into lakes and ponds, including Dachan Lake, Magu

沉湖朝霞　李梓固摄
Morning Glow of Chenhu Lake　Photo by Li Zigu

Lake and Taibai Lake, leading to the disappearance of the "Yunmeng Wetland" in historical records. During the Southern Dynasties, the largest lakes were Dachan and Magu , of which it was said that "in summer, the two lakes contain as much water as the sea". Taibai Lake was located in the south of today's Caidian District, where the present-day Chenhu Lake is found. It was mentioned in the book *Notes on the Book of Waterways*.

到了唐代，大浐、太白二湖不再见于记载，马骨湖的面积和深度也远不及南朝。到宋代，马骨湖也不见记载，成为葭苇弥望的沼泽地，有"百里荒"之称。

In the Tang Dynasty, there was no record of Dachan and Taibai, and the area and depth of Magu Lake were far diminished in comparison with the Southern Dynasties. In the Song Dynasty, records of Magu Lake, too, ceased, and the area it had occupied changed into large reed-covered marshes known as the "One Hundred-li Wasteland".

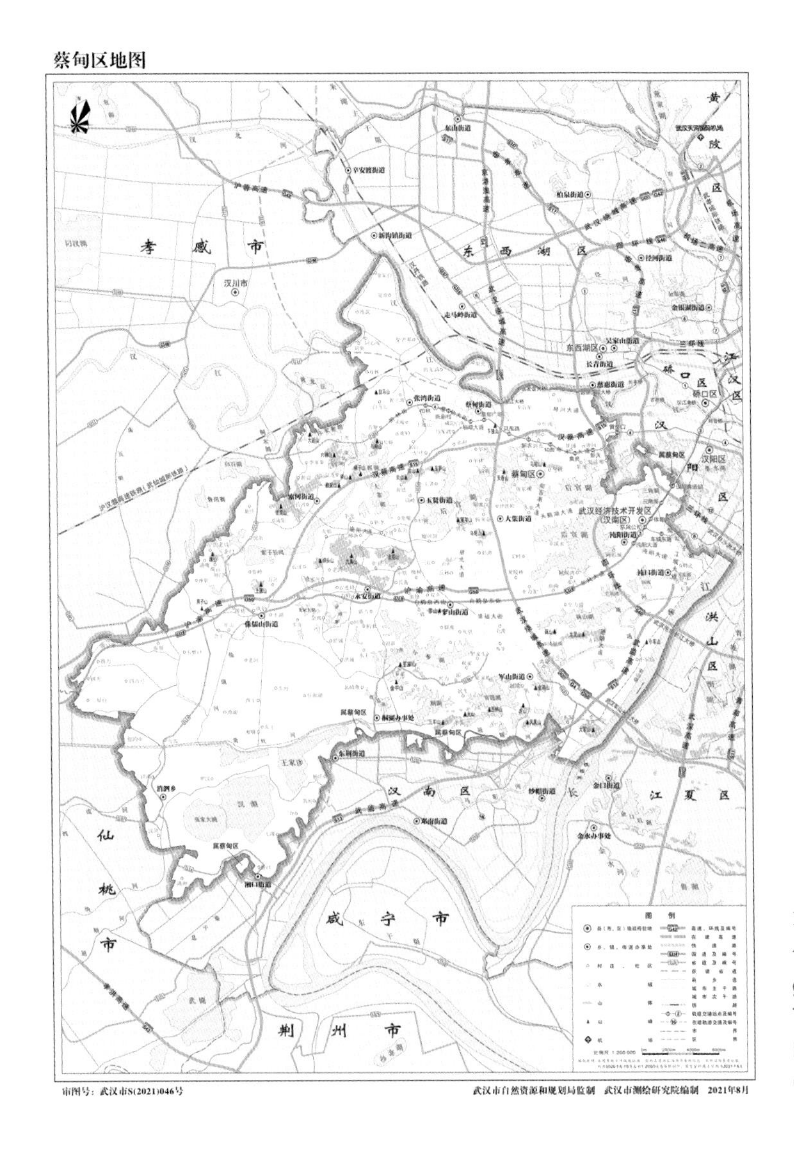

蔡甸区地图
供图：天地图（www.tianditu.gov.cn）
The Map of Caidian District
Photo Credit: Map World (www.tianditu.gov.cn)

宋以后，原云梦泽地区湖泊水面缩小的趋势没有继续。根据明清两代的史料和舆图记载，这一地区湖泊数量众多，又出现了太白湖。春夏水涨时与附近其他小湖连成一片，成为江汉间众水所归的“巨浸”。清乾隆时期，太白湖改称赤野湖。到百余年后的光绪年间，太白湖又基本消失了。直到现在，江汉平原上仍星罗棋布地分布着被称作“鄂渚”的二百多个浅小的湖泊，正是古云梦泽被分割、解体而残留的遗迹。

After the Song Dynasty, the former water area of the Yunmeng Wetland ceased to shrink. Historical records and maps from the Ming and Qing Dynasties show that there were many lakes in the area. Taibai Lake reappeared, and connected with other small lakes in the vicinity when its water level rose in spring and summer, forming a large lake, into which all water bodies between the Han River and the Yangtze flowed. During Emperor Qianlong's reign in the Qing Dynasty, Taibai Lake was renamed Chiye Lake. Over one hundred years later, during Guangxu's reign, Taibai Lake almost disappered again. To this day, existing as the divided remnants of the ancient Yunmeng Wetland, over 200 small and shallow lakes called "Ezhu" are still widely scattered over the Jianghan Plain.

新中国成立后，各级政府在江汉平原泛水区进行了修堤筑垸、围湖造田等综合治理，各湖沼逐步成为封闭型水体，水陆面积趋向稳定。该区域是我国重要的粮食生产基地和经济发达地区，在我国空间经济布局上有重要地位。

After the founding of the People's Republic of China in 1949, the government at all levels conducted comprehensive management in the flooding area of the Jianghan Plain, including the construction of dikes and embankments, and the land reclamation, rendered each lake a relatively closed body of water with a stable water area, and stabilized the surrounding land areas. Now being important for food production and (being) economically developed, this region of China occupies an important position in the nation's arrangement of spatial economy.

2 福荫江城
A Protector of Wuhan

沉湖湿地位于武汉市蔡甸区西南部，长江中下游江汉平原东部长江与汉水汇流的三角地带，北接索子长河直达汉水，南隔东荆河遥望汉南，东邻四湖（独沧湖、小奓湖、桐湖、官莲湖），西毗仙桃。

The Chenhu Lake Wetland in the southwest of Caidian District, Wuhan City, is located in the delta where the Han River and the Yangtze merge. It is in the east of the Jianghan Plain in the middle and lower reaches of the Yangtze. It is connected to the Han River farther north by the Suozichang River, adjacent to the Dongjing River to the south, overlooking Hannan District, next to the four lakes (Ducang Lake, Xiaozha Lake, Tonghu Lake, and Guanlian Lake) to the east, and Xiantao City to the west.

晚霞中的沉湖　李梓固摄
Chenhu Lake at Sunset　Photo by Li Zigu

沉湖由长江、汉江泛溢淤积而成，位于汉水与长江漫滩交汇而构成的低洼地段，除北部部分地区为波状平原外，其余皆为平原。地面高程为海拔 17.5m ~21m，由多个碟形洼地复合构成。最低处为沉湖，并以此为中心向四周辐射缓升，形如炊锅。沉湖保护区总面积 115.8 km^2，湿地面积为 67.5 km^2，主要由王家涉湖、沉湖和张家大湖构成。三湖紧紧相连，仅一堤之隔。湖区主要河流为东荆河，即通顺河，古时称沌水，在沌口注入长江。西侧黄丝河是东荆河的一条主要支流，从仙桃市东流进沉湖，并与汉江的杜家台分洪道融为一体，经王家寨闸汇入东荆河。

Chenhu Lake lies in a low-lying area formed at the intersection of the Han River and the Yangtze by their flooding and sedimentation. Apart from some undulating parts in the north, it is a relatively flat plain at an elevation of 17.5−21 meters above sea level, containing several saucer-shaped depressions. With Chenhu Lake at its lowest point, its elevation gradually increases from the center to the periphery, with a profile curved like a wok. With a total area of 115.8 square kilometers, the Chenhu Lake Wetland Nature Reserve contains a wetland area of 67.5 square kilometers composed mainly of Wangjiashe Lake, Chenhu Lake and Zhangjia Lake, which are closely connected, separated only by a single dike. The main river in the lake area is the Dongjing River (Tongshun River), known as the Zhuan River in ancient times, which merges into the Yangtze River at Zhuankou. On the west side is the Huangsi River (a major tributary of the Dongjing River), which flows east from Xiantao City into Chenhu Lake to join the Han River's Dujiatai Floodway, and merges into the Dongjing River via Wangjiazhai Water Gate.

对于武汉而言，沉湖不仅带来了水草丰美、鸟飞鱼跃的极致美景，成为武汉人民休闲观光的“后花园”，更重要的是，沉湖湿地有巨大的调蓄洪水、净化水质、生物多样性维持等生态功能，为武汉市经济发展和人民生活提供了良好的环境和资源，在城市生态建设和经济社会可持续发展中有着不可替代的重要作用，可谓江城武汉的“忠诚卫士”。

The Chenhu Lake Wetland is very important to Wuhan, for it brings out ultimate beauty of many waters, lush grass, flying birds and jumping fish, and offers Wuhan citizens a “backyard garden” for leisure and sightseeing. More importantly, with great ecological functions such as flood regulation and storage, water purification, and biodiversity maintenance, the Chenhu Lake Wetland provides an excellent environment and ample resources to support economic development and improve the lives of the people of Wuhan. It plays an irreplaceable role in urban ecological construction and sustainable economic and social development. Hence it has the title of “Loyal Guard” of Wuhan, the City of Rivers.

“防洪卫士”调蓄洪水
“Flood Defender”: Flood Regulation and Storage

汉江是长江中下游最大支流。汉江中上游河谷开阔，水面较宽，泄洪能力较强，而下游河道上宽下窄呈“漏斗形”，河宽从 600m~2500m 递次缩窄至 200m~300m，泄洪能力上大下小，极不平衡，导致汉江中下游地区历史上洪涝灾害频繁而严重。杜家台分洪工程是汉江下游唯一的分洪工程，由汉江进洪闸（杜家台分洪闸）、行洪道、蓄洪区（汉南泛区）和长江泄洪闸（黄陵矶闸）等部分组成。行洪道与东荆河为原汉江天然分流故道，汉南泛区历史上曾是长江和汉江的自然洪泛区，沉湖是杜家台分洪工程蓄洪区的一部分，在汛期分蓄汉江洪水，减轻洪水对武汉市的威胁。经统计，汛期分洪蓄水可降低洪峰水位 0.6m~3.0m，为保障汉江下游和武汉市的防洪安全发挥了巨大作用，防洪效益十分显著。

The Han River is the largest tributary in the middle and lower reaches of the Yangtze. After the river flows through the open valleys in its middle and upper reaches, its water surface becomes wider, boasting a strong capacity for discharging floods. However, in its lower reaches, the river has a "funnel shaped" profile, with a wide top and a narrow bottom, and gradually narrows from 600–2500 meters to 200–300 meters. Thus the Han River's capacity for discharging floods is stronger in its upper reaches but far weaker in its lower reaches. This imbalance has historically caused frequent and severe floods in its middle and lower reaches. The sole flood diversion project in the lower reaches of the Han River is the Dujiatai Flood Diversion Project, composed of an intake gate (the Dujiatai Flood Diversion Gate) on the Han River, a spillway, a flood storage area (the Hannan Floodplain), and a flood discharge gate (the Huanglingji Gate) on the Yangtze River. The spillway and the Dongjing River were natural diversion channels for the Han River, while the Hannan Floodplain has historically served as a natural flood overflow area for both the Yangtze and the Han River. Chenhu Lake serves as part of the Dujiatai Flood Diversion Project's flood storage area, responsible for the diversion and storage of the Han River floodwater during the flood season in order to reduce the threat of flooding to Wuhan City. Statistically, it can lower the flood level by 0.6–3.0 meters of flood diversion storage during the flood season, playing a great role in supporting the safe flood control in Wuhan City and the lower reaches of the Han River. Its performance in flood control is impressive and remarkable.

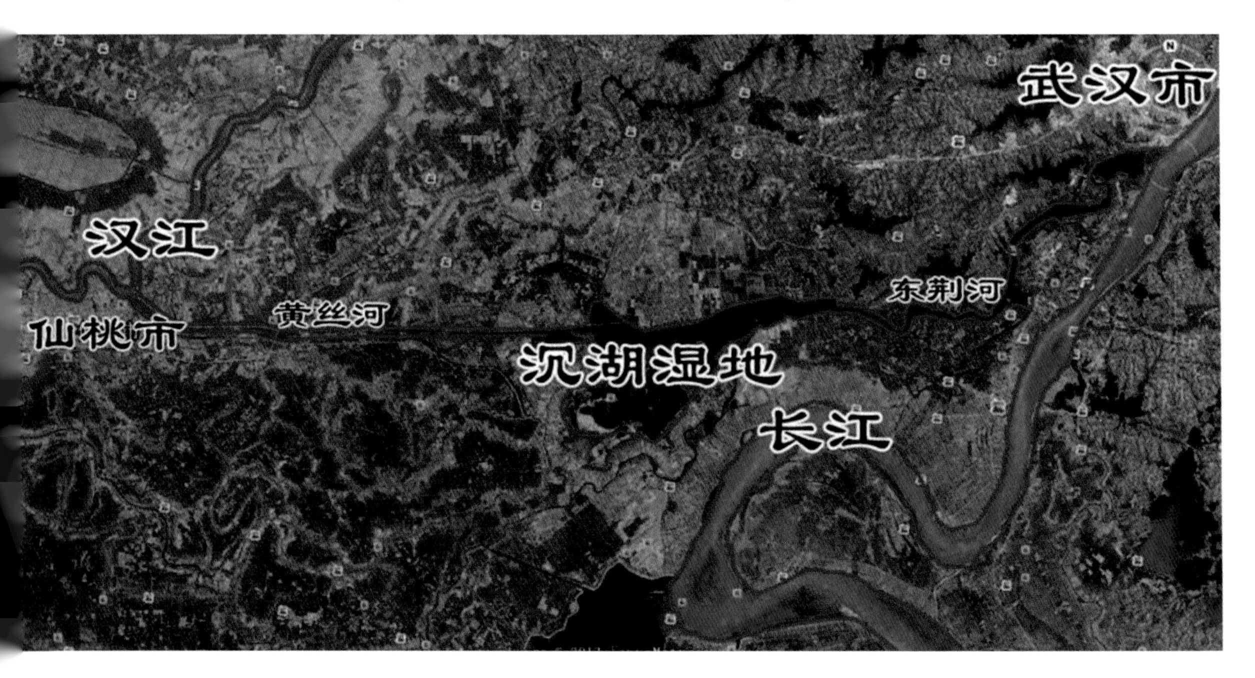

春到沉湖，绿草茵茵　魏斌摄
Lush Grass in the Spring of Chenhu Lake　Photo by Wei Bin

“城市之肾”污染治理：
The “City’s Kidney”: Pollution Management

武汉地处长江中游，汉江下游，受上游下泄的污染较为严重。沉湖湿地具有净化水质的功能，可谓“城市之肾”。

Located in the middle reaches of the Yangtze and the lower reaches of the Han River, Wuhan is subject to serious pollution due to upstream discharges. With its water purifying capability, the Chenhu Lake Wetland can be called the “City’s Kidney”.

沉湖的水生植物群落从岸边到湖心随水位变化可分为四个植物带：湿生植物带—挺水植物带—浮叶植物带—沉水植物带。湿生植物带以红穗薹草、水田碎米荠等为主。挺水植物带以芦苇、荻、水烛、香蒲和菰为主。浮叶植物带以菱、四角菱、芡实为主。沉水植物带以竹叶眼子菜、菹草、狐尾藻和金鱼藻为主。由于湿地植物的拦截作用，

当水流进入湿地后，速度明显降低。较慢的水流速度有助于沉积物的下沉，也有助于与沉积物结合在一起的营养物和悬浮物的储存和转换。各种植物、微生物通过复杂的物理、生物、化学过程，吸收、分解、转化污染物和有毒物质。植物组织中富集重金属的浓度比周围水中浓度高出 10 万倍以上。

Chenhu Lake's aquatic plant community can be classified into four plant zones ranging from its banks to its center, in terms of changes in water level: the zones for hygrophyte, emergent plant, floating leaf plant, and submerged plant. The main species in the hygrophyte zone are *Carex argyi*, *Cardamine lyrata*, etc. The main emergent plants include *Phragmites communis, Triarrhena sacchariflora, Typha angustifolia*, *Typha orientalis*, and *Zizania*

小天鹅和村庄　魏斌摄
The Village and Tundra Swans　Photo by Wei Bin

latifolia. The main species growing in the floating leaf plant zone include *Trapa bispinosa, Trapa quadrispinosa*, and *Euryale ferox, etc. Potamogeton wrightii, Potamogeton crispus, Myriophyllum verticillatum* and *Ceratophyllum demersum* are growing in the submerged plant zone. These wetland plants interfere with the flow of water, significantly decreasing its velocity after it enters the wetland. This slower velocity in turn allows sediment to sink, contributing to the storage and conversion of nutrients and suspended solids which are intermixed with sediment. Thus the various plants and microorganisms here absorb, decompose and transform pollutants and toxic substances through complex physical, biological and chemical processes, resulting in concentrations of heavy metals in local plant tissues over 100,000 times higher than those in the surrounding water.

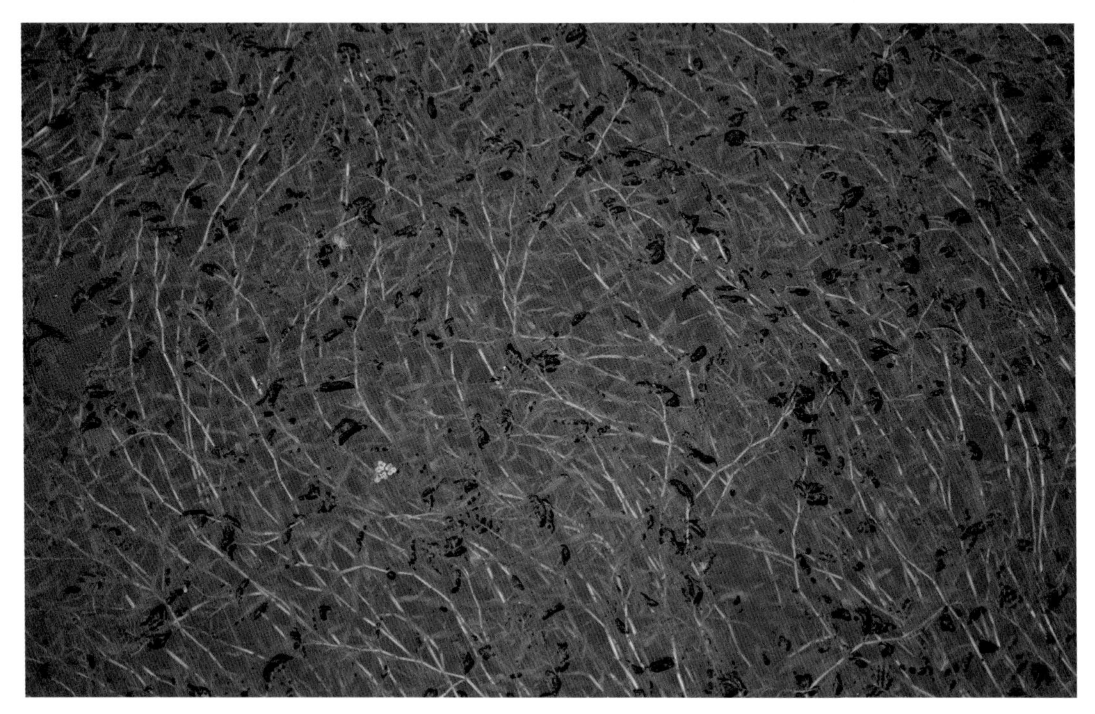

菹草　孔雪摄
Potamogeton crispus　Photo by Kong Xue

菱　供图：沉湖湿地省级自然保护区管理局
Trapa bispinosa　Photo Credit: Chenhu Lake Wetland Provincial Nature Reserve Administration

近年来，武汉市地表水水质稳步提升。2021 年，166 个湖泊中，劣Ⅴ类水质湖泊全面清零，实现了历史性突破。作为蔡甸区与武汉经济技术开发区（汉南区）的跨区水域接点，老关村提取的水质监测指标还被用作上游的考核断面和对下游的对照断面。如此精准考核，也使得长江武汉段水质连续优于国家考核目标。

In recent years, Wuhan's surface water quality has been steadily improving. In 2021, in an historic breakthrough, all 166 lakes were removed from the inferior Class V ranking. At Laoguan Village, a cross-regional water connection point between Caidian District and Wuhan Economic & Technological Development Zone (Hannan District), the water quality monitoring indicators provide cross-sectional assessment data for the upstream reaches and cross-sectional control data for the downstream reaches. Such precise assessment is a further reason for the water quality in the Yangtze River's Wuhan section continuously exceeding national assessment targets.

“生态家园”生物多样性维持
“Eco Home”: Biodiversity Maintenance

沉湖地区气候条件良好，土质肥沃，入湖径流营养物质多，营养盐类含量高、底泥厚且营养丰富，极利于水生生物生长，水生生物资源极其丰富，是莲藕、芡实、菱等水生植物的主要生长区。

Since the Chenhu Lake area offers good climate and fertile soil, and its water carries rich nutrients, contains nutrient salts and has thick, nutrient-rich substrate sludge at the bottom, it is extremely conducive to the growth of aquatic organisms. These extremely abundant aquatic biological resources have made Chenhu Lake a major area for the growth of aquatic plants, such as lotus, gorgon fruit, and *Trapa bispinosa*.

每年夏季，东荆河流域 3000 km^2 的径流汇入湖区，加上长江涨水通过黄陵闸倒灌，沉湖地区的沟渠河道便和湖区连成一片汪洋。而在秋冬季及早春枯水季节，湖水排江，余水落湖，只有湖心、渠道还保持着不到 1m 深的水面，仅占丰水期湖水面积的 15%，其余部分则形成了大片的泥泞草甸，构成了浅湖—沼泽—草甸连续的湿地生态系统，为越冬水鸟提供了优良的栖息环境，是众多珍稀水鸟重要的越冬地和迁徙停歇地。红穗薹

湖泊 滩涂 冬候鸟 魏斌摄
Winter Birds on the Lake and Marsh Photo by Wei Bin

草集中分布于沉湖、张家大湖、王家涉湖等湖泊及黄丝河滩涂，生物量最大，为湿地鸟类提供了丰富的食物来源。汛期水淹之后，这里又是鲤和鲫的产卵、索饵及避敌场所。

Every summer, as the runoffs from the Dongjing River Basin with an area of 3,000 square kilometers flow into the lake area, and the Yangtze River rises and flows backwards through the Huangling Gate, the canals and riverways surrounding the Chenhu Lake area merge into a broad expanse of water. In autumn and winter, as well as early spring, excess water from the lake is discharged into the river. While some water is left in the lake, only its central area and the canals contain water less than one meter deep, accounting for only 15% of the lake area of its wet period; the rest of the lake area is covered with large muddy

meadows. Thus, at this time, a continuous wetland ecosystem integrating shallow lakes, marshes and meadows is established in this area, providing an excellent habitat for wintering waterfowl and making Chenhu Lake an important wintering ground and resting place for many rare migratory waterbirds. *Carex argyi*, a species of sedge, grows most densely in

沉湖滩涂，飞翔的灰雁群　魏斌摄
A Flock of Greylag Geese over the Marshes of Chenhu Lake　Photo by Wei Bin

Chenhu Lake, Zhangjia Lake, Wangjiashe Lake and other lakes, as well as on the Huangsi River banks. This plant provides a rich food source for wetland birds. Once the area floods during the flooding season, it offers habitat for carp and crucian carp to spawn, feed and conceal themselves from predators.

红穗薹草　供图：沉湖湿地省级自然保护区管理局
Carex argyi　Photo Credit: Chenhu Lake Wetland Provincial Nature Reserve Administration

湿地水文变化带来的洲滩季节性积水，也为钉螺提供了适宜的生存环境，加之众多的人口和不合理的开发利用，血吸虫治理一直是一项重要工作。历史上，长江中下游地区的血吸虫病人占全国血吸虫病人的60%以上。以往稻田有一定水位和湿润泥滩，可满足雁类等水鸟的栖息。为了改变钉螺孳生环境、加强血吸虫病防治管理，在湖北省农业产业结构调整的背景下，周边很多地区将水稻改为大棚蔬菜、油菜、棉花等旱地作物。沉湖更是成了水鸟不可或缺的栖息地。

The control of disease-causing parasitic schistosomes here is a tricky mission, as the seasonal accumulation of water on the banks offers habitat suitable for *Oncomelania hupensis*, due to the wetland's hydrological changes, and its large population and over-exploitation. Historically, the middle and lower reaches of the Yangtze River possessed over 60% of China's schistosomiasis patients. In the past, the specific water levels and wet mudflats of paddy fields created a habitat particularly suiting waterbirds such as ducks, geese and swans. To disrupt the breeding environment for *Oncomelania hupensis* and improve schistosomiasis control, many areas surrounding Chenhu Lake have taken advantage of Hubei Province's agricultural restructuring policy, and switched from rice cultivation to greenhouse vegetables, oilseed rape, cotton and other upland crops, leaving Chenhu Lake as an indispensable habitat for waterbirds.

沉湖生态系统完整，生态功能特殊，野生动植物资源十分丰富，是长江中游地区最具典型意义的湖泊湿地之一。沉湖湿地于 1994 年建立了区级自然保护区，次年晋升为市级自然保护区，2006 年又成为武汉市第一个省级湿地自然保护区。先后列入《中国重要湿地名录》和《亚洲湿地名录》。2000 年被列入《中国湿地保护行动计划》，2004 年被列入《中国湿地保护工程 2005—2010 年规划》。2007 年，沉湖湿地被列为长江中下游湿地保护网络首批重要成员。2009 年国际鸟盟将沉湖湿地列为国际重要鸟区，2013 年 10 月，沉湖湿地列入国际重要湿地名录。

反嘴鹬集群和民房　魏斌摄
A Flock of Avocets in Front of Houses　Photo by Wei Bin

With its complete ecosystem, special ecological functions and rich wildlife resources, the Chenhu Lake Wetland is a typical one of the lake wetlands in the Yangtze River' s middle reaches. The district-level nature reserve established there in 1994 was upgraded to municipal-level nature reserve status in 1995, and then becoming the first provincial-level wetland nature reserve in Wuhan in 2006, and being included first in the *List of China's Important Wetlands* and then the *List of Wetlands in Asia.* Added to the *China National Wetland Conservation Action Plan* in 2000 and the *2005–2010 Plan for National Wetland Conservation Program* in 2004, it was included in the first batch of important members of the Yangtze River' s middle and lower reaches wetland conservation network in 2007. In 2009, Bird Life International listed the Chenhu Lake Wetland as a bird habitat of international significance; in October 2013, Chenhu Lake was included in the *List of Wetlands of International Importance.*

第二章

Chapter II

大城绿水

A Great City with Clear Water

湖泊的形成、演变过程，往往与城市的发展有着密切关系。由于人类亲水的天性与对水的需求，滨水区往往成为城市发展的起点，同时也是受人类活动影响最强烈的地带。在武汉这一特大城市的西南隅，依然保留着自然山水的野趣。沉湖如一颗碧绿的明珠，波光粼粼，熠熠生辉。作为《湿地公约》第十四届缔约方大会参观考察点，向全世界展现着一幅人与自然和谐发展的“中国绿”画卷。

The gradual formation of lakes is often closely related to the development of cities. Due to humanity's water-loving nature and need for water, lakeside areas often serve as a starting point for urban development, and so these areas are also most strongly influenced by human activities. Thus, the Chenhu Lake Wetland, a bright green pearl in the southwest of the megacity of Wuhan, continues to retain the natural landscape's rustic charm, glittering and glistening. As a visit and observation point for the 14th Meeting of the Conference of the Contracting Parties to the *Ramsar Convention on Wetlands* (COP14), this area also presents a "Green China" picture of harmonious development of humanity and nature to the world.

 武汉唯一的国际重要湿地
Wuhan's Only Wetland of International Importance

166 个湖泊，165 条河流。星罗棋布的湖泊与纵横阡陌的水网，散落交织在辽阔的江汉平原之上，演绎出大武汉的万千气象，构成典型的“水乡泽国”和“鱼米之乡”的景观。在这众多的湿地中，沉湖是唯一的国际重要湿地。

Wuhan boasts 166 lakes and 165 rivers. These numerous lakes and the network of vertical and horizontal water channels are scattered and intertwined across the Jianghan Plain, reflecting Wuhan’s magnificence and presenting scenes such as “Land of Rivers and Lakes” and “Land of Fish and Rice”. Among the many wetlands in Wuhan, the Chenhu Lake Wetland is the only one of international importance.

根据《湿地公约》规定，满足 9 项标准中的任意一项，即可申报国际重要湿地。而沉湖湿地满足其中的 5 项：

According to the *Convention on Wetlands*, application for wetlands of international importance is allowed as long as any one of the nine criteria is met, and the Chenhu Lake Wetland meets the following five criteria:

标准 1：如果一个湿地包含适当生物地理区内一个自然或近自然湿地类型的一处具代表性的、稀有的或独特的范例，就应被认为具有国际重要意义。

Criterion 1: A wetland should be considered internationally important if it contains a representative, rare, or unique example of a natural or near-natural wetland type found within the appropriate biogeographic region.

长江中下游地区湿地分布广泛且类型多样，是我国淡水湖泊分布最集中和最具有代表性的地区，也是世界湿地和生物多样性保护的热点区域。沉湖是长江中下游湿地中极典型的淡水湖泊和泛水沼泽湿地。

With diverse wetlands widely distributed in the middle and lower reaches of the Yangtze River, this region becomes the most representative one with the densest concentration of freshwater lakes in China. It also is a hot spot region for wetland and biodiversity conservation in the world. Chenhu Lake is a very typical freshwater lake and flooded marsh wetland in the middle and lower reaches of the Yangtze River.

标准 2：如果一个湿地支持着易危、濒危或极度濒危的物种或者受威胁的生态群落，就应被认为具有国际重要意义。

Criterion 2: A wetland should be considered internationally important if it supports vulnerable, endangered, or critically endangered species or threatened ecological communities.

沉湖湿地支持着 IUCN 红色名录多种受胁鸟类在此栖息，包括白鹤、青头潜鸭、黄胸鹀等。

The Chenhu Lake Wetland supports many species of birds on the *IUCN Red List of Threatened Species*, including Siberian cranes, Baer's pochards, and yellow-breasted buntings.

白鹤优雅挺立　魏斌摄
A Siberian Crane Standing Gracefully　Photo by Wei Bin

标准 4：如果一个湿地在生命周期的某一关键阶段支持物种或在不利条件下对其提供庇护场所，就应被认为具有国际重要意义。

Criterion 4: A wetland should be considered internationally important if it supports plant and/or animal species at a critical stage in their life cycles, or provides refuge during adverse conditions.

沉湖位于东亚—澳大利西亚迁飞区，生境丰富多样，在 270 余种鸟类的生活史中发挥着重要作用。沉湖是雁鸭类候鸟重要的越冬地及鸻鹬类等候鸟的重要中转站，还是部分候鸟的重要繁殖地。

Located on the East Asia-Australasia Flyway, Chenhu Lake offers rich and diverse habitats, playing an important role in the life history of over 270 species of birds. It is an important wintering ground for geese and ducks, a major transit station for plovers and sandpipers and other migratory birds, and a significant breeding ground for some migratory birds.

标准 5：如果一个湿地定期栖息有 2 万只或更多的水鸟，就应被认为具有国际重要意义。

Criterion 5: A wetland should be considered internationally important if it regularly supports 20,000 or more waterbirds.

反嘴鹬，群浪　魏斌摄
A Wave of Avocets　Photo by Wei Bin

沉湖湿地每年都有超过 2 万只的水鸟在此栖息。在 2019 年、2020 年、2021 年越冬季，最高记录分别达到 60003、78789、86642 只。

Over 20,000 waterbirds inhabit the Chenhu Lake Wetland every year. The record-high numbers of birds reached 60,003, 78,789, and 86,642 in the wintering periods of 2019, 2020, and 2021 respectively.

标准 6：如果一个湿地定期栖息有一个物种或亚种某一种群 1% 的个体，就应被认为具有国际重要意义。

Criterion 6: A wetland should be considered internationally important if it regularly supports 1% of the individuals in a population of one species or subspecies of waterbird.

根据 2007~2021 年沉湖湿地鸟类调查监测结果统计，沉湖湿地有豆雁、灰雁、罗纹鸭、灰鹤、普通鸬鹚、白琵鹭等 18 个物种的数量超过了其全球种群数量的 1%。

According to the survey and monitoring results about birds in the Chenhu Lake Wetland during 2007–2021，the numbers of 18 species of birds, including bean goose, greylag goose, falcated duck, common crane, common cormorant, Eurasian spoonbill, etc., in this wetland exceeded 1% of their total numbers in the world.

CONVENTION ON WETLANDS

(Ramsar, Iran, 1971)

《湿地公约》：为保护全球湿地及湿地资源，1971 年 2 月 2 日，来自 18 个国家的代表在伊朗小镇拉姆萨尔共同签署了《关于特别是作为水禽栖息地的国际重要湿地公约》(简称《湿地公约》，又称《拉姆萨尔公约》)。《湿地公约》是全球第一个政府间的多边环境公约，也是唯一以一种类型生态系统为对象的公约。《湿地公约》早于《物种迁徙公约》《濒危物种贸易公约》《气候变化公约》《生物多样性公约》等，其宗旨是通过地区和国家层面的行动及国际合作，推动全球湿地的保护管理与合理利用，以此为实现全球可持续发展做

出贡献。当前《湿地公约》已有 171 个缔约方，是签约国家最多的环境公约。我国于 1992 年加入《湿地公约》。

The *Convention on Wetlands*: To protect wetlands and wetland resources worldwide, representatives from 18 countries signed the *Convention on Wetlands of International Importance Especially as Waterfowl Habitat* (The *Convention on Wetlands*, also known as the *Ramsar Convention*) in the Iranian town of Ramsar on February 2, 1971. *The Ramsar Convention* is the world's first intergovernmental treaty that provides a multilateral framework for environmental conservation and is the only convention targeting a single type

清晨，白琵鹭集群觅食飞翔　魏斌摄
Eurasian Spoonbills Flying and Hunting for Food at Dawn　Photo by Wei Bin

of ecosystem. *The Convention on Wetlands* was issued before the *Convention on Migratory Species,* the *Convention on International Trade in Endangered Species of Wild Fauna and Flora,* the *United Nations Framework Convention on Climate Change,* the *Convention on Biological Diversity*, etc. Its purpose is to promote the conservation, management and rational use of wetlands worldwide through regional and national actions and international cooperation, thereby contributing to global sustainable development. Currently, there are 171 parties to the *Convention on Wetlands*, which is the environmental protection convention involving the largest number of contracting countries. China joined this convention in 1992.

2017 年，《湿地公约》决定在全球范围内启动国际湿地城市认证工作，将国际湿地城市作为推广《湿地公约》保护目标和方法的样板。获得认证代表一个城市对湿地生态保护的最高成就，是一块厚重的国际“金字招牌”，同时将有助于提升城市的综合竞争力，增加国际美誉度，提升城市整体形象，促进区域绿色发展。武汉市正在积极申报“国际湿地城市”这一殊荣，沉湖湿地成为资源本底项评选标准中 1+6 块“重要湿地”的“1”。

In 2017，the *Convention on Wetlands* decided to launch the international wetland city certification globally and used the International Wetland Cities as models to promote its conservation objectives and methods. This certification is to commend a city for its highest achievement in wetland ecological protection and serve as a significant international “prize”. It also will help enhance the city’s comprehensive competitiveness, increase its international reputation, improve the overall image of the city and promote its regional green development. Wuhan is actively applying for the honorary title “International Wetland City”. The Chenhu Lake Wetland has become the “one” important wetland in the “1 + 6 important wetlands” evaluation references for the resource background projects.

2 沉湖四季　沉醉忘归
The Chenhu Lake Wetland’s Entrancing Seasons

春季百花盛开、草长莺飞；夏季波光粼粼、碧荷千顷；秋季天清气朗、芦花摇曳；冬季万鸟齐飞、诗情碧霄。四时之景不同，而乐亦无穷也。在这个鱼翔雁栖之湖，人与自然和谐相处的美妙画卷，正徐徐展开。

At the Chenhu Lake Wetland, visitors can see flowers blooming and grass flourishing, and hear birds singing in spring. In summer, the sparkling lakes are enveloped by vast expanses of lotus flowers. In autumn, the reed flowers sway under the clear sky. In winter, thousands of birds fly across the blue sky in poetic visions. The changes of Chenhu Lake’s scenery in the four seasons offer visitors endless enjoyment. This lake, as a habitat for numerous fish and harmonious coexistence of humanity and nature.

白鹤双飞　魏斌摄
A Pair of Siberian Cranes Flying Together　Photo by Wei Bin

沉湖风光：飞翔的小天鹅和湖面的冬候鸟　魏斌摄
Flying Tundra Swans and Other Winter Birds over Chenhu Lake　Photo by Wei Bin

沉湖王家涉湖，滩涂和冬候鸟群　魏斌摄
Flocks of Winter Birds on Wangjiashe Lake of the Chenhu Lake Wentland　Photo by Wei Bin

卷羽鹈鹕，两只共同觅食　魏斌摄
A Pair of Dalmatian Pelicans Hunting for Food　Photo by Wei Bin

湿地春和·草木蓬勃

Grass and Trees Greening Up in Springtime

和着春雷，早春时节，沉湖大片浅滩上呈现出“草色遥看近却无”的景致，大地露出微微笑意。至春和景明，湖面沙鸥翔集，锦鳞游泳；岸边草甸郁郁葱葱，如厚织绿毯，野花次第盛放，流连其间，戏蝶时舞、野蜂忙碌。

灰雁降落　魏斌摄
Greylag Geese Landing　Photo by Wei Bin

阳春三月，白琵鹭和绿头鸭　魏斌摄
Eurasian Spoonbills and Mallards in Early Spring　Photo by Wei Bin

沉湖湿地俯瞰　颜军摄
A Bird's Eye View of the Chenhu Lake Wetland　Photo by Yan Jun

As early spring's thunder rumbles, Chenhu Lake's vast shallow shores begin to take on the fascination of "green grass best perceived from afar". It seems that the water is smiling. In the warmth of spring, gulls soar above the lake in flocks, schools of fish swim in the lake, the meadows on the shore are lush, green and thick as carpets, the wildflowers are blooming with bees busy shuttling between them, and the butterflies are dancing.

大地像个信手点染的画家，嘟囔着："饱和度再高点，再高点……"金色的油菜花海冲击着来者的视线。仲春者，浴乎无垠春光，风乎馥郁芬芳，咏而归。岂不快哉？

The earth is like a skilled painter constantly muttering "brighter color!" A sea of golden oilseed blossoms gives visitors a visual shock. In the second month of spring, outings to beautiful scenes of flowers and grass provide unsurpassed enjoyment and satisfactory homecomings.

湿地夏景·辽原旷野

Boundless Wilderness in Summer

夏季江河丰盈，远望水天辽阔，横无际涯；近观荷叶田田，荷花婀娜。泛舟湖上，置身四面芙蓉，找寻乡野喜乐。软风拂面，香满江湖烟水，不禁赞叹“惟有绿荷红菡萏，卷舒开合任天真”。

In summer, the rising rivers and distant sky appear large in the beholders’ eyes. The lotus leaves and flowers gracefully extend, and visitors can immerse themselves amongst them, seeking rustic pleasure by boating in the center of the lake. With soft breezes caressing their faces, the visitors cannot help but exclaim, “The green of the lotus leaves and the red of its flowers match perfectly.”

湿地秋色·斑斓飞羽

The Landscape in Autumn

沉湖的秋斑斓而澄净、空灵而丰富，辽阔、蛮荒、带着野性。天地玄黄，灰蓝相渲，

沉湖湿地怒放的野荷 李梓固摄
Wild Lotus Growing in the Chenhu Lake Wetland Photo by Li Zigu

张家大湖　颜军摄
Zhangjia Lake　Photo by Yan Jun

棉凫降落　颜军摄
Cotton Pygmy Geese Landing　Photo by Yan Jun

茫茫苇海漫无际涯，浅滩、沼泽、湖泊皆隐逸其间。摇曳的芦秆绵延数里，似浪花涌动，连素穗，翻秋气，细节疏茎任长吹。

The Chenhu Lake Wetland nowadays is a vast area of colorful, diversified, wild, and ethereal scenery. The sky is deep blue, and the water is bright clear. The sea of reeds is stretching for miles, sheltering shallow shores, marshes and lakes, swaying like waves, nodding in the autumn wind.

蒹葭苍苍、白露为霜之际，数以万计的候鸟归来，诗意地践行着生命回归的承诺。鸻鹬仰俯、雁鸭泅凫、鸥鹭忘机、渊邈天鹅舞。

In "White Dew", the 15th solar term of the Chinese calendar, thousands of migratory birds return to stay here, fulfilling their poetic promise of returning to life. Plovers, geese, ducks, gulls, herons and swans are living freely in the wetland.

湿地冬日 · 飞鸟翩跹

Flying Birds in Winter

冬季的沉湖，在幽静与闲适中休养生息。烟波浩渺的湖面，成群鸟儿或伫立或信步，或游弋或起落，又增几许热闹与繁忙。

沉湖风光，苇荡和远处的鸟群 魏斌摄
Reed Marshes and Bird Flocks of Chenhu Lake Photo by Wei Bin

苍鹭　颜军摄
A Grey Heron Standing　Photo by Yan Jun

普通鸬鹚群　魏斌摄
Common Cormorants　Photo by Wei Bin

反嘴鹬群　魏斌摄
A Flock of Avocets　Photo by Wei Bin

小天鹅飞行　颜军摄
Tundra Swans Flying　Photo by Yan Jun

Chenhu Lake is the ideal place for birds to spend a quiet, leisurely winter, and visitors can see flocks of birds standing or walking, landing or swimming on the misty surface of the lake, making a bustling scene.

几只白鹤从水墨画里飞起，轻点水面；惊鸿一瞥，大红鹳又来栖居；还有鹬类组成的庞大鸟浪，在空中翻腾、变幻，蔚为壮观。

A few Siberian cranes fly in the sky, which is like an ink painting, to lightly skim the water surface. People may catch a glimpse of the flamingos that inhabit this lake. Huge waves of shorebirds, somersaulting and changing their formation high in the sky, may give you a spectacular view.

湖区居民“靠湖吃湖”，民间称这片养育他们的湖区为“三来宝地”。所谓“三来”，就是“芦苇湖草土中蹦出来，各种鱼群长江中游进来，大雁野鸭天上飞过来”。长期的湖区生产生活，使“沉湖人”形成了一些具有地方特色的生态产品和民俗文化。

东方白鹳——国家一级保护动物　雷刚摄
The Oriental Stork, a National First-Class Protected Animal　Photo by Lei Gang

白鹤　魏斌摄
Siberian Cranes　Photo by Wei Bin

反嘴鹬　颜军摄
The Pied Avocet　Photo by Yan Jun

白琵鹭，觅食和降落　魏斌摄
Eurasian Spoonbills Landing for Food　Photo by Wei Bin

白鹤、灰鹤、小天鹅等　魏斌摄
Siberian Cranes, Common Cranes, Tundra Swans, etc.　Photo by Wei Bin

黑腹滨鹬群飞　颜军摄
A Flock of Dunlins Flying　Photo by Yan Junn

The residents in the lake area make good use of their lake resources for a living. Locals call this lake area the "Treasured Place of Three Sources". The so-called "Three Sources" means that their living materials come from the grass and soil surrounding the reed-covered lake, from all kinds of fish swimming in from the Yangtze River, and from geese and wild ducks flying in the sky. During their long-term production life in the lake area, locals have developed some ecological products with local characteristics and folk culture.

“芦”柴

Reeds

湖区居民将湖中数万亩的芦苇荡称为“柴山”。虽说叫“山”，却是一马平川，好像草原一样。过去，立冬前后芦柴成熟，泛水退出，收割芦苇的时候也就到了。收割工人先在“柴山”里选一块地势较高的地方，割去芦苇，然后用捆好的芦苇为墙，中间留 6 尺左右的空当，搭成一个长方形的棚子。棚子周围挖排水沟，人就住在棚中。这种芦棚不漏雨、不漏雪，还很保暖。割芦柴的人脚踏一寸多厚的木拖鞋，以防芦桩刺破脚板。割下的芦柴一般五六米长，再用芦苇剖成篾扎捆。“柴山”里生长的芦柴有三种，一种叫芦苇，用于打芦席；一种叫刚柴，用于编帘子；一种叫毛柴，主要用于做燃料。现在，“柴山”归蔡甸区芦苇场管理，一年可产芦柴 3 万吨左右，主要用作造纸原料。

The residents in the lake area used to call the reed marsh covering an area of tens of thousands of *mu* (*mu*, a Chinese unit for measuring area) in the lake “Firewood Hill”, where they cut reeds as the firewood they needed . Although called Hill, the marsh is actually as flat as a prairie. In the past, when reeds were ripe around the Start of Winter, the 19th solar term of the Chinese calendar, the floodwater would recede from the Hill, and it was time to harvest reeds. Harvesters first chose a higher position, cut some reeds, and placed

王家涉湖 颜军摄
Wangjiashe Lake Photo by Yan Jun

the bundled reeds opposite each other. A 6-*chi* (*chi*, a Chinese length unit) gap should be left between the bundles of reeds to set up a rectangular shed, around which a drainage ditch would be dug. The shed served as a shelter for the harvesters. This kind of reed shed could prevent rain or snow from entering and keep people warm. While cutting the reeds, the workers wore their wooden slippers with a one-*cun*-plus thick sole (*cun*, a Chinese length unit) to prevent reed pile tips from piercing their feet. The cut reeds, usually five or six meters long, were bundled with the rind of other reeds. There are three kinds of reeds growing on the Firewood Hill. The first type is called *luwei* (*Phragmites australis*), which is used to weave reed mats; the second is called *gangchai* (*Triarrhena lutarioriparia*), which is used to weave curtains; the third is called *maochai* (*Imperata cylindrica*), which is mainly used as fuel. At present, the Firewood Hill is under the management of the reed yard in Caidian District. The reeds now are mainly used for paper making. Chenhu Lake can produce about 30,000 tons of reeds per year.

“沉”鱼
Fish

由于湖内饵料充足，水面开阔，沉湖中的鱼类生长速度快。有经验的渔民说，沉湖鱼喜欢打雷天，每经历一次风雨，这里的鱼就会长一尾长。泛水季节，渔民用“花篮”捕鱼。编“花篮”，是先用细篾编扎成两个周围有孔的半圆柱体，再用薄篾将两个半圆柱连起来，中间留一空处不连，以作取鱼之口。“花篮”两端各留一口，另用细篾做成呈漏斗状排列的“倒挂须”，这样鱼就只能进而不能出。将“花篮”放入二三尺深鱼常出没的水中，鱼见竹篾映水有花，便入内玩耍，不知不觉陷身笼中。第二天，渔民用前端有钩的长竿提起“花篮”，这样捕的鱼比其他任何方法都鲜活。

As there is abundant fish food in Chenhu Lake and its water surface is wide, fish in the lake grow quite fast. According to experienced fishermen, fish in Chenhu Lake like thundering days, and every time they experience a storm, the fish here will grow by the length of their tails. During the flooding season, fishermen tend to use “flower baskets” to catch fish. To weave such a flower basket, the first step is to weave two hollow half-cylinders with fine bamboo

沉湖，冬捕的渔民 魏斌摄
Fishing on Chenhu Lake in Winter Photo by Wei Bin

strips. The second step is to use thin bamboo strips to connect these two half-cylinders. Between the two half-cylinders should be left an opening to fetch fish. Another two openings should be left on both ends of the flower basket, and into which is tied something funnel-shaped called "Dao Gua Xu" with thin bamboo strips. This design ensures that fish can only enter, but cannot go out. When the flower basket is placed two or three *chi* deep under the water where fish usually live, the fish see the "flowers" made with bamboo strips reflected in the water, enter the basket to play, and get trapped in the cage unknowingly. The next day, the fisherman will use a long rod with a hook at the tip to lift the flower basket. The fish caught in this way is fresher than those caught in any other method.

深秋时节，沉湖开始退水，这时也到了“放鱼”季节。所谓“放鱼”，不是把鱼放走，而是将水放走，把鱼留下。先在泄水的地方建柴帘闸口，使鱼不得随水而走脱，然后在出水口安装一长龙圆网，称之为“毫”，用粗绳将“毫”尾扎住，放水滤鱼。水放到一定程度，便见成群结队的鱼铺天盖地向下游涌去，这场景真是难以形容。这时有经验的“毫师傅”赶紧把“毫”尾打开倒翻，将四角固定在木桩上，成为一个方形“井口”，鱼随水进“毫”，“毫师父”则轮班握着捞子不停地往船上舀鱼。有时一天要出几万斤鱼。

Late autumn is a season for the water of Chenhu Lake to recede, and also the season to "release fish", which means to drain water away for keeping fish. The first step is to use a reed-made curtain gate at the water drainage site to ensure that fish will not flee away with the flowing water. The next step is to install a long net, called "Hao", at the lake outlet, and use a thick rope to tie its tail for filtering water to keep fish. After the water is drained to a certain extent, it can be seen that swarms of fish rush downstream. The scene is indescribable. At this time, experienced "Hao" masters will uncover the tail end of "Hao" and turn it upside down to fix the four corners to the piles. Thus, a square opening can be formed to make fish enter "Hao" with water. These masters often take shifts to scoop the fish with a fishing tool onto the boats. Sometimes, they can fetch tens of thousands of *jin* （a Chinese weight unit） of fish.

沉湖鳙鱼由于其独特的地域环境和“人放天养”的养殖模式，形成了独特的品质。这里的鳙鱼头大尾小、体形肥硕，鳞片紧密、光洁鲜亮，肉质细嫩、味道鲜美，比其他同等体长的鱼重 10%~20%。高品质的渔获引来无数客户青睐，不仅在武汉地区占有一席之地，还远销港澳台地区，成为沉湖水产养殖的重要品牌。

Due to the unique geographical environment and breeding mode of "human rearing plus natural breeding", the quality of bighead carps in Chenhu Lake is distinctive. Bighead carps produced here have big heads, small tails, fat bodies, tight scales and a bright body surface. The fish meat is tender and tasty. Such carps are 10% – 20% heavier than other fish of the same size. Drawing favor from numerous customers, this high-quality fish has secured

沉湖鱼汤　供图：沉湖湿地省级自然保护区管理局
Fish Soup Photo Credit: Chenhu Lake Wetland Provincial Nature Reserve Administration

an important status in the Wuhan market. It has been an important brand of local aquaculture, and also sold to Hong Kong, Macao and Taiwan of China.

“藕”遇
Lotus Roots

蔡甸素有“中国莲藕第一乡”的美誉。这里出产的莲藕不仅通长肥硕、质细白嫩、藕丝绵长，而且口味香甜、生脆少渣、极富营养。清澈的湖水、肥沃的湖泥，加之汉江多次改道，洪水泛滥带来的丰富矿物质淤积，为蔡甸莲藕的生长提供了物质基础。蔡甸植藕历史悠久。隋唐时期，人工栽培的莲藕开始在蔡甸传播引种，良种沃土相得益彰，越长越好。宋代，蔡甸莲藕开始闻名京都，蔡甸莲花湖莲藕年年入贡。明清时，蔡甸已经是大面积植藕。清代诗人乔大鸿在《晚渡南湖》诗中写道：“十里尽薄荷，迷漫失南湖。人家何处边，停桡问渔叟。”蔡甸莲藕自 2007 年开始获得地理标志产品保护。

沉湖采藕人　供图：沉湖湿地省级自然保护区管理局
Lotus Roots Harvesting from Chenhu Lake　Photo Credit: Chenhu Lake Wetland Provincial Nature Reserve Administration

沉湖藕汤　供图：沉湖湿地省级自然保护区管理局 The Soup of Lotus Roots Photo Credit: Chenhu Lake Wetland Provincial Nature Reserve Administration

Caidian District has long been reputed as "China's First Town of Lotus Roots". The lotus roots produced here are long and fat, with fine, white and tender flesh, long fibers and a sweet taste. They are crisp, low-residue and extremely nutritious. The clear water and fertile mud of the lake and deposition of minerals caused by flooding and repeated diversion of the Han River offer a sound material basis for the growth of lotus roots in Caidian. Caidian has a long history of lotus root planting. During the Sui and Tang dynasties, the artificially cultivated lotus roots were introduced in Caidian as a cultivated species. Due to the perfect combination of good seeds and fertile soil, the lotus roots here grew better and better. Since the Song Dynasty, the lotus roots of Caidian became famous in the capital. The lotus roots from the Lotus Lake in Caidian were selected as a tribute and transported to the capital every year. In the Ming and Qing dynasties, lotus roots were planted on a large scale, the evidence of which can be found in the poem entitled *"Crossing South Lake in the Evening"* by Qiao Dahong, a poet of the Qing Dynasty. The poem says, "Lotus leaves and flowers are scattered miles, and I am boating in the South Lake but lost, I ask the fisherman where the residential dwellings are." Since 2007, Caidian's lotus roots have been protected as a product with national geographical indication.

每年中南海国宴，蔡甸莲藕从不缺席。藕和肉骨煨汤，文火煨到肉烂脱骨，藕炖得软糯粉嫩，油水清亮。一口咬下去十几根藕丝不断，越拉越长。以莲藕为主题的饮食文化丰富多彩，莲藕入菜五花八门：藕圆、藕夹、滑藕片、炝拌藕丁、财鱼焖藕、清炒藕带、油炸荷花、冰镇莲米、荷叶蒸肉、银耳莲子羹……俱为舌尖上的美味，也是江湖生活里最熨帖的人情味和世代难忘的记忆。

Every year, Caidian's lotus roots are available at the state banquets at Zhongnanhai. They are often used to make soup with meat-carrying bones. The cooking method is to simmer the meat on a small fire until it falls off the bones; to stew the lotus roots until they are soft and tender; to cook the soup until the oil and water turn clear. If you bite the lotus root, tens of lotus root fibers will extend longer and longer, without being bitten off. Due to the district's diversified food culture in the theme of "lotus roots", lotus roots have been used in a wide variety of dishes, such as Lotus Root Powder Ball, Fried Lotus Root Sandwich, Stir-fried Lotus Roots, Stir-fried Diced Lotus Roots, Stewed Snake-headed Fish with Lotus Roots, Stir-fried Lotus Sprouts, Fried Lotus Blossoms, Iced Lotus Seeds, Steamed Pork Wrapped in Lotus Leaves, White Fungus Soup with Lotus Seeds... These dishes are all the delicacies of Caidian people, representing the most thoughtful feelings in their daily life and the memorable memories of the previous generations.

第三章

Chapter III

候鸟福地

An Ideal Habitat for Migratory Birds

鹤鸣雁飞、花叶葳蕤，万物生灵栉风沐雨，人与自然美美与共，丰富绚烂的生物多样性是城市可持续发展的基石。

沉湖是东亚—澳大利西亚迁飞区关键的候鸟栖息地，也是长江中下游湿地保护网络的重要一环。沉湖虽面积不算大，但珍稀水鸟物种多、种群数量大，在长江中下游的湖泊中独具特点。不断打破纪录的鸟类物种数和罕见鸟种记录，使沉湖成为武汉最为耀眼的明星湿地。每年冬季，数万只候鸟聚会于此，体现了湿地的健康程度与超强的承载能力；罕见鸟种频频现身，彰显了沉湖近乎完美的生物多样性。

沉湖目前记录有鸟类 277 种，包括湿地水鸟 127 种，其中近一半为冬候鸟，包括东方白鹳、黑鹳、卷羽鹈鹕、白鹤、青头潜鸭、黄胸鹀等珍稀物种。

In a world that values coexistence of humanity and nature, and pursues biodiversity which is the cornerstone of sustainable urban development, all beings can survive and prosper.

Chenhu Lake is a key migratory bird habitat on the East Asia-Australasia Flyway and an important element of the wetland conservation network in the Yangtze River's middle and lower reaches. Chenhu Lake is not very big, but it boasts rare waterfowl species with large populations, making it distinctive. Constantly breaking records for bird species numbers and rare bird species, Chenhu Lake has become Wuhan's most brilliant wetland. Every winter, there are tens of thousands of migrants gathering here, reflecting Chenhu Lake's health and huge carrying capacity. Rare species which are frequently witnessed highlight the lake's near-perfect biodiversity.

Of the 277 bird species recorded on Chenhu Lake, 127 are waterfowl species of which nearly half are winter migrants, including the Oriental stork, black stork, Dalmatian pelican, Siberian crane, Baer's pochard, and yellow-breasted bunting.

滩涂、芦苇荡、冬候鸟　魏斌摄
Marshes, Reeds, and Winter Birds　Photo by Wei Bin

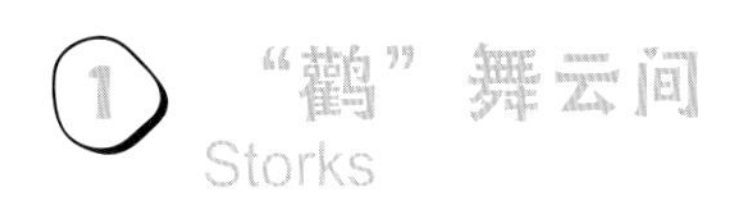

1 "鹳"舞云间 Storks

东方白鹳
Oriental Stork

鹳是古老的物种，曾在地球上广泛分布。目前，全世界共有 6 属 19 种鹳，中国记录有其中 7 种：东方白鹳、白鹳、钳嘴鹳、彩鹳、秃鹳、白颈鹳、黑鹳。

Storks are ancient birds that were once widely distributed on the earth. Currently there are 19 species in 6 genera worldwide. China has records of seven stork species: the Oriental stork, white stork, Asian openbill, painted stork, marabou, woolly-necked stork and black stork.

东方白鹳，是鹤形目鹳科鹳属的大型涉禽，仅分布在东亚地区，主要繁殖地为我国东北地区和俄罗斯远东地区，越冬地为我国长江中下游地区。20 世纪中叶以来，东方白鹳种群数量急剧下降，相继在日本、朝鲜半岛野外灭绝。在中国，黑龙江三江平原和嫩江中下游地区是东方白鹳的重要繁殖地。近年来，在东方白鹳的迁徙停歇地陆续发现繁殖巢，在越冬地繁殖的记录也逐渐增加。据 1995~2005 年全球水鸟种群数量估计，东方白鹳全球数量约 3000 只。近年来野生种群数量估计增至 7000~9000 只。

东方白鹳　雷刚摄
The Oriental Stork　Photo by Lei Gang

东方白鹳在 IUCN 濒危物种红色名录中列为濒危，在 CITES 公约中列入附录 I。2021 年 2 月 1 日，在国家林业和草原局、农业农村部公告发布的《国家重点保护野生动物名录》中，东方白鹳作为新增保护物种，被列为国家一级重点保护野生动物。这次名录级别调整对东方白鹳的保护意义重大。

The Oriental stork (*Ciconia boyciana*), a large wading bird belonging to the genus Ciconia, family Ciconiidae and order Ciconiiformes, is found only in East Asia. Breeding primarily in the Russian Far East and northeastern China, it overwinters in the middle and lower reaches of China' s Yangtze River. Since the middle of the 20th century, the Oriental stork population has declined sharply, resulting in its extinction in the wild in Japan and on the Korean Peninsula. In China, Oriental storks mainly breed on the Sanjiang Plain and in the middle and lower reaches of the Nenjiang River. In recent years, Oriental stork nests have been found at their stopover sites, and Oriental storks also increasingly breed in their wintering grounds. According to the global waterbird population estimates for 1995–2005, the global population of the Oriental storks was around 3,000. However, in recent years, their global population has increased to 7000–9000. Listed as endangered on the *IUCN Red List of Threatened Species*, the Oriental stork is also included in Appendix I of the *Convention on International Trade in Endangered Species* (CITES). On February 1, 2021, it was made a first-class protected species via being added to the *List of State Key Protected Wild Animals* announced by the National Forestry and Grassland Administration and the Ministry of Agriculture and Rural Affairs of China. This is of great significance to its protection.

东方白鹳眼周有一小圈红色裸皮，虹膜呈浅黄色，黑色的嘴长而粗壮。鹳类没有完善的鸣管和发达的鸣肌，很少鸣叫，而是通过上下喙急速叩击发出“哒哒”声，进行个体间信息交流。东方白鹳多在 10cm~30cm 的浅水区觅食，取食鱼类、贝类、甲壳类以及植物种子、根及嫩叶等。

The Oriental stork has bare red skin around its eyes, pale yellow irises, and a long, stout black bill. Lacking well-developed syrinx and syringeal musculature, the stork's calling is muted; instead, they communicate via bill-clattering. The Oriental stork mainly feeds in shallow water 10–30 centimeters deep. The diet of the Oriental stork consists of fish, shellfish, crustaceans, as well as plant seeds, roots and young leaves.

东方白鹳和大白鹭群飞　魏斌摄
Oriental Storks Mixed Up with Great Egrets　Photo by Wei Bin

20世纪80年代以来，东方白鹳几乎每年都造访沉湖，曾是沉湖的“招牌”。东方白鹳每年10月下旬至11月上旬迁到沉湖，翌年3月下旬至四月上旬迁离。1985年，武汉大学胡鸿兴教授观测到有近420只。1988年1月27日原武汉市汉阳县人民政府发布了《关于严禁捕猎白鹳等珍稀水禽的布告》；1989年3月1日，《中华人民共和国野生动物保护法》颁布实施；同年7月22日，《武汉市野生动物保护规定》发布施行，这些法律和法规的颁布实施，对东方白鹳、白琵鹭和鸳鸯等珍禽的保护发挥了重要作用。1991年2月，国际鹤类基金会主席詹姆斯•哈里斯先生也专程到沉湖考察湿地水禽及其生态环境，认为“沉湖湿地保持着较好的自然景观，对鹤类、鹳类以及其他水禽的越冬具有极为重要的价值”。此后，哈里斯先生曾4次给湖北省政府及省野生动物主管部门的领导致函，希望尽快在沉湖建立湿地水禽自然保护区。1994年，沉湖越冬东方白鹳种群达到历史最高纪录900余只后，数量不断减少，至2004年种群数量逐渐减少到了100只以下，直到2011年，一只也未出现。近年来，东方白鹳偶尔在沉湖短暂停留。2021年冬天，阔别多年的东方白鹳终于再次在沉湖越冬，12月21日，王家涉湖记录到了38只东方白鹳聚群，为近20年来历史最高纪录。

The Oriental storks have visited Chenhu Lake almost every year since its first visit recorded in the 1980s, becoming the lake's "signature". Migrating to Chenhu Lake each year between late October and early November, they leave between late March and early April the next year. In 1985, Professor Hu Hongxing of Wuhan University observed nearly 420 Oriental storks. On January 27, 1988, the government of Hanyang County, Wuhan City, issued the *Notice on Prohibiting Hunting Oriental Storks and other Rare Waterfowl*; on March 1, 1989, the *Law of the People's Republic of China on the Protection of Wildlife* was promulgated and implemented; and on July 22, 1989, the *Regulations of Wuhan City on Wildlife Protection* was put into effect. The implementation of these laws and regulations has played an important role in the protection of rare birds such as the Oriental stork, Eurasian spoonbill, and mandarin duck. In February 1991, Mr. James Harris, President of the International Crane Foundation, paid a special visit to Chenhu Lake to inspect the waterfowl and ecological environment, and considered that the wetland had retained a good natural landscape, being of great significance for overwintering cranes, storks and other waterfowl. After his visit, Mr. Harris wrote four times to the officials of Hubei Provincial Goverment and the wildlife authority of Hubei Province, calling for the establishment of a waterfowl nature reserve in Chenhu Lake as soon as possible. After reaching a record high of over 900 in 1994 the wintering population of Oriental storks began to decline, falling to less than 100 in 2004, with no Oriental stork over-wintering in 2011. In recent years, the birds stayed at Chenhu Lake for only a few days, until 2021, when the Oriental storks finally wintered at Chenhu Lake again after years of absence, with 38 Oriental storks recorded in Wangjiashe Lake on December 21, a 20 year record high.

黑鹳

Black Stork

黑鹳身体主要颜色为黑色，喙、眼周的裸皮区域和腿部为鲜艳的红色，腹部为白色，古人称之为“玄鹤”。黑鹳的分布地几乎贯穿欧亚大陆北部，我国北方很多地区都有它们的分布，多为留鸟。部分黑鹳种群秋季会迁徙到华中、华南越冬。

黑鹳——国家一级保护动物　供图：沉湖湿地省级自然保护区管理局
The Black Stork, a National First-Class Protected Animal　Photo Credit: Chenhu Lake Wetland Provincial Nature Reserve Administration

黑鹳歇息　魏斌摄
Black Storks at Rest　Photo by Wei Bin

The black stork (*Ciconia nigra*) has a white belly, with bright red skin on its bill and legs, and around its eyes, was known as “the black crane” in ancient times because of its black body. Black storks are distributed throughout most of northern Eurasia and also found in many areas in northern China. Most of them are non-migratory, but in autumn, some migrate to winter in central and southern China.

2021 年 11 月 16 日，22 只黑鹳聚群现身王家涉湖，是进入 21 世纪来最大种群数量。黑鹳每年可见，大群首次出现。此后，一个 5 只的小团队在沉湖越冬，这是 2010 年记录到 9 只以后的最大种群，往年一般只有 1 到 2 只。

Black storks are observed every year. On November 16, 2021, 22 black storks were observed together at Wangjiashe Lake. This was the first time they were in such a large group, the largest in this new century. Then, five black storks wintcred at Chenhu Lake, the largest population since nine were recorded in 2010. Five is a remarkable number compared to the one or two typically observed in previous years.

2 “鹤”鸣九皋 Cranes

“鹤鸣于九皋，声闻于野。”这是距今两三千年前，先人对鹤类在沼泽地栖息的生动描述。鹤科是鸟类中一个古老的科，出现的时间比人类要早 6000 多万年。几

灰鹤降落　魏斌摄
Common Cranes Landing　Photo by Wei Bin

灰鹤　颜军摄
Common Cranes　Photo by Yan Jun

千万年以前，地球上曾经存在 30 多种鹤类，但很多只在地层中留下了化石。现存鹤类共 15 种，我国有 9 种，是世界上拥有鹤类种群数量最多的国家。丹顶鹤、白鹤、赤颈鹤、黑颈鹤、白头鹤为国家一级保护动物，灰鹤、沙丘鹤、白枕鹤、蓑羽鹤为国家二级保护动物。

"The crane cries in the marsh, and her cries are heard in the open country." The verse from two or three thousand years ago vividly described the scene of a crane resting by the marsh. Cranes are an ancient family that appeared more than 60 million years earlier than humans. Tens of millions of years ago, there were more than 30 crane species on the earth, but many of them have become fossils. Of the 15 extant crane species, 9 are found in China, the country with the largest crane population. The red-crowned crane, Siberian crane, sarus crane, black-necked crane and hooded crane are listed in China as national first-class protected animals, while the common crane, sandhill crane, white-naped crane and demoiselle crane are listed as national second-class protected animals.

鹤是所有飞行鸟类中身材最高的，其高雅秀逸的身姿，深得中国人民和世界人民喜爱。在中华文化中，鹤美丽、飘逸的形象和长寿、吉祥的寓意已经成为人们审美和高洁人文精神的象征，如明清时期一品文官服才有仙鹤图案。松乃百木之长，鹤为百羽之宗，国人素有以“松鹤图”贺寿之习俗。

Cranes, the tallest flying birds, are favored by people in China and around the world for their elegant and graceful postures. Cranes represent beauty, elegance, longevity and good luck in Chinese culture, and have become a symbol of aesthetics and humane spirit. For example, on the uniform of first-rank civil officials in the Ming and Qing dynasties was a crane badge. Believing that pines are the king of all trees, and cranes the queen of all birds, the Chinese have developed the custom of giving a pine-and-crane painting as a birthday gift to convey good wishes.

鹤类的安宁和昌盛，也是世界人民和平相处的象征。中国于 1997 年加入了东北亚鹤类网络，并于 1999 年与俄罗斯等 9 个国家共同签署了《关于白鹤保护措施的谅解备忘录》。近年来，鹤类见证了中国的湿地恢复。据中国野生动物保护协会鹤类联

灰鹤大群　颜军摄
A Flock of Common Cranes　Photo by Yan Jun

合保护委员会2021年开展的调查，目前，中国黑颈鹤、丹顶鹤、灰鹤的分布区基本保持稳定，白鹤和白头鹤的分布区有向周边扩大的趋势。从种群数量来看，中国黑颈鹤的越冬种群数量显著增长，白鹤的种群数量略有增加，灰鹤的种群数量相对稳定。

Being composed and vigorous, the crane is also a symbol of peaceful coexistence of people around the world. After joining the North East Asian Crane Site Network (NEACSN) in 1997, China signed the *Memorandum of Understanding Concerning Conservation Measures for the Siberian Crane* with Russia and eight other countries in 1999. In recent years, cranes have witnessed the restoration of China's wetlands. According to a survey in 2021 by the Joint Crane Protection Committee of the China Wildlife Conservation Association, the distribution areas of black-necked cranes, red-crowned cranes and common cranes in China have basically retained their sizes, while those of Siberian cranes and hooded cranes have expanded. In terms of population size, black-necked cranes have increased significantly, Siberian cranes have increased slightly, and common cranes have remained relatively stable.

呼朋引伴的白鹤
Siberian Crane

扫码观赏视频
Scan and watch

白鹤是国家一级重点保护野生动物，IUCN红色名录极度濒危物种。白鹤可以分为东部、中部和西部种群，都在俄罗斯西伯利亚北部繁殖，分别在中国、印度和伊朗

白鹤　颜军摄
Siberian Cranes　Photo by Yan Jun

白鹤群飞　魏斌摄
Siberian Cranes Flying in a Flock　Photo by Wei Bin

越冬。其中，东部种群每年迁徙于俄罗斯西伯利亚与中国长江中下游地区之间，是数量最多的白鹤种群。

The Siberian crane (*Grus leucogeranus*), a first-class national key protected wild animal of China, is listed as a Critically Endangered species in the *IUCN Red List*. The species is divided into eastern, central and western populations that all breed in northern Siberia in Russia and winter in China, India, and Iran respectively. The eastern population, being the largest in size, migrates annually between Siberia and the middle and lower reaches of China's Yangtze River.

湖泊水位下降以后形成的大面积草滩和浅水湖泊是白鹤的主要生境。白鹤喜食莲藕的嫩尖，经常在有莲分布的区域盘旋，寻找脆嫩美味。往年在沉湖有2~3只白鹤稳定越冬，以家庭为聚群，分散活动。2020年冬季最多纪录到9只。2021年冬，白鹤、灰鹤、白枕鹤、蓑羽鹤和白头鹤在沉湖济济一堂，其中白鹤达到13只。武汉记录过的鹤类这个冬季首次在沉湖聚齐。

Siberian cranes mainly inhabit large grassland and shallow water formed by the decline of the water level of lakes. They often hover over lotus flowers, looking for their favorite food—the crisp and tender tips of lotus roots. In former years, there were two or three Siberian cranes overwintering at Chenhu Lake gathering as family flocks and acting separately. The wintering population was up to nine in 2020. In the winter of 2021, Chenhu Lake witnessed the first gathering of all crane populations recorded in Wuhan, including the Siberian crane, common crane, white-naped crane, demoiselle crane, and hooded crane, among which the wintering population of Siberian cranes reached 13.

灰鹤

Common Crane

灰鹤广泛分布于欧亚大陆，是在我国境内越冬地分布最广的鹤。北至辽宁、南及海南、西到云南，都有灰鹤分布，而其主要的越冬地在黄河和长江周边的湿地。根据历年监测，灰鹤一般在沿岸滩涂和农田活动，主要食物为草根、谷粒、麦苗，也吃小型螺、蚌等。

灰鹤水面飞翔　魏斌摄
A Common Crane Flying above the Water　Photo by Wei Bin

Distributed across Eurasia, the common cranes (*Grus grus*) cover the most widespread wintering grounds in China, ranging from Liaoning in the north through Hainan in the south to Yunnan in the west. They mainly overwinter in the wetlands near the Yellow River and the Yangtze River. According to monitoring data, common cranes are often found in waterfront mudflats and farmlands, and feed mainly on grass roots, grains, wheat seedlings, small snails, mussels, etc.

灰鹤携幼飞翔 魏斌摄
A Common Crane Flying with Younger Ones Photo by Wei Bin

沉湖是灰鹤北迁的中转站。在上个越冬季，沉湖最多记录到4000多只灰鹤。2022年2月24日，武汉刚转晴，上万只灰鹤从空中分批降落沉湖，煞是壮观。休整一夜后，它们在清晨启程北飞。

Chenhu Lake is the "transit point" for common cranes moving northwards. Up to 4,000 common cranes were recorded last winter at Chenhu Lake. On February 24, 2022, when

灰鹤集群，晨光中觅食 魏斌摄
Common Cranes Hunting for Food at Dawn Photo by Wei Bin

Wuhan just turned sunny, more than 10,000 common cranes landed in batches, forming quite a spectacular view. After a night off, they set off to the north in the morning.

3 水暖“鸭”知 Ducks

鸭类是常见的迁徙水鸟，发达的胸肌和修长的翅膀，是它们长途迁徙的保障，带蹼的脚掌使它们善于游泳，有的鸭类还善于潜水。鸭类有尾脂腺，能够分泌足够的油脂，以便羽毛防水。

Ducks are common migratory waterfowl, with well-developed pectoral muscles and slender wings for long-distance migration, and webbed paws for swimming. Some species are good divers. Ducks have uropygial glands that can produce oil to make their feathers water-repellent.

赤麻鸭　颜军摄
A Ruddy Shelduck　Photo by Yan Jun

沉湖有 20 余种鸭类，其中绿头鸭、斑嘴鸭为留鸟，棉凫在此繁殖，其余均为冬候鸟。沉湖罗纹鸭、赤麻鸭、绿翅鸭、青头潜鸭的数量均超过了其全球种群数量的 1%。历年来沉湖越冬鸭类无论是从种类和数量上均呈逐渐上升的趋势。生态恢复后的 1800 亩荷塘，是鸭类理想的聚集地。

绿头鸭降落　颜军摄
A Mallard Landing　Photo by Yan Jun

There are more than 20 duck species in Chenhu Lake. Among them, mallards and spot-billed ducks are permanent residents, cotton pygmy geese breed here, and the rest are all winter migrants. The populations of falcated ducks, ruddy shelducks, green-winged ducks and Baer's pochards in Chenhu Lake exceed 1% of their global populations. Over the past years, ducks overwintering at Chenhu Lake have been on the rise in both the number of species and size of populations. The restored 1,800 *mu* of lotus pond provides an ideal gathering place for ducks.

青头潜鸭
Baer's pochard

扫码观赏视频
Scan and watch

青头潜鸭，是我国一级重点保护野生动物。青头潜鸭曾是广泛分布于我国的重要鸟类之一，然而自 20 世纪 80 年代末期以来，其全球及我国国内种群数量急剧下降。我国种群从 1987~1993 年的 16700 只锐减到 2003~2011 年的 2131 只；全球种群从 10000~25000 只锐减到 5000 只，繁殖栖息地与越冬栖息地也不断退缩。2008 年，青头潜鸭被 IUCN 定为濒危物种，由于其濒危现状并未得到改善，其种群数量与分布记录仍持续减少。2013 年，青头潜鸭被升级为极度濒危物种。目前，青头潜鸭的全球种群数量被认为不足 1000 只。

青头潜鸭，雄鸭理羽　魏斌摄
A Male Baer's Pochard Preening Itself　Photo by Wei Bin

青头潜鸭，雌鸭飞行　魏斌摄
A Flying Female Baer's Pochard Photo by Wei Bin

Baer's pochard (*Aythya baeri*) is a first-class national protected wild animal that was once widely distributed in China. However, since the late 1980s, its population has declined dramatically both in China and abroad. Its number in China had undergone sharp decline from 16,700 in 1987–1993 to 2,131 in 2003–2011, and from 10,000–25,000 to 5,000 worldwide. Its breeding and wintering grounds are also shrinking. In 2008, Baer's pochard was classified as Endangered (EN) by IUCN, and in 2013, Critically Endangered (CR), as the situation did not improve and the population continued to fall, with a global population of less than 1,000.

青头潜鸭为一种中等体型鸭类。成年雄鸟繁殖羽头部暗绿色，阳光下有金属光泽，虹膜白色，喙铅灰色，胸部红棕色，胁部清晰的白色区域带有棕色条带。青头潜鸭雌鸟体型与雄鸟相似，在各年龄段、各季节虹膜均为暗色，无白色“眼眶”。青头潜鸭分布于亚洲东部、东南部，是东亚—澳大利西亚迁飞区的重要物种。主要在俄罗斯远东地区与中国东部繁殖，在中国南部、缅甸、泰国、孟加拉、印度、日本等地越冬。

Baer's pochard is a medium-sized duck. Adult males have a dark-green head that looks metallic in daylight. Their irises are white, bill lead gray, and breast reddish-brown, and there are brown stripes on their white flanks. Female Baer's pochards are similar in size to males but with dark irises at all ages and in all seasons and without white feathers around their eyes. Distributed in eastern and southeastern Asia, Baer's pochard is an important species on the East Asia-Australasia Flyway, breeding primarily in the Russian Far East and northeastern China, overwintering in regions and countries including southern China, Myanmar, Thailand, Bangladesh, India, Japan, etc.

青头潜鸭繁殖栖息地为多水生植物的湖泊、河流和池塘，常在河湖岸边的草丛中或灌丛下营巢，有时为浮巢；冬季多栖息在大型湖泊、江河、海湾、河口、水塘和沿海沼泽地带。近年来，青头潜鸭在其越冬地长江中下游也出现了繁殖记录。青头潜鸭通常成对或成小群，在水边水生植物丛中或附近水面上活动，不喜欢水流湍急的河流，越冬及春秋迁徙会集结十几只或上百只的群体，有时也会与凤头潜鸭或其他潜鸭混群栖息。

Baer's pochards mainly inhabit and breed in lakes, rivers and ponds with rich aquatic vegetation. They often builds nests on a tussock or under shrubs along the banks, sometimes on the water. In winter, they mainly inhabit large lakes, rivers, bays, estuaries, ponds, and coastal marshes. In recent years, their breeding has also occurred in wintering grounds in the middle and lower reaches of the Yangtze River. Baer's pochards often appear in pairs or small groups in waterfront aquatic plants or on nearby water, and are hardly seen on fast-flowing rivers. They gather in groups of a dozen or a hundred during wintering, and spring and autumn migration. Sometimes they also share habitats with tufted ducks or other diving ducks.

青头潜鸭主要以各种水草的根、叶、茎和种子等为食，也吃软体动物、水生昆虫、甲壳类、蛙等动物性食物。觅食方式主要通过潜水，但也能在水边浅水处直接伸头摄食。

Feeding mainly on water plant roots, leaves, stems and seeds, the diet of Baer's pochards also includes mollusks, aquatic insects, crustaceans, frogs and other animals. The duck feeds mainly by diving, or just extends its head to get food in shallow waters.

2013 年，在武汉黄陂府河一带首次发现青头潜鸭。2014 年，10 余只青头潜鸭首次被观察记录到在府河湿地筑巢繁殖。2017 年 12 月，武汉市观鸟协会成员在黄陂区祁家湾发现 263 只的大型青头潜鸭种群。近年来，它们在武汉不断“扩圈”，武汉市观鸟协会会员又在牛山湖、上涉湖等地也发现了它们的栖息地，证明了武汉已成为青头潜鸭重要的栖息地、繁殖地。

In 2013, Baer's pochard was first discovered along the Fuhe River in Huangpi District, Wuhan. In 2014, a dozen Baer's pochards were first observed nesting and breeding in the Fuhe River wetland. In December 2017, members of the Wuhan Birdwatching Association found a large flock of 263 Baer's pochards in Qijiawan, Huangpi District. As more and more ducks of this kind are present in Wuhan in recent years, members of the Wuhan Birdwatching Association have also found its nests in different places including Niushan Lake and Shangshe Lake, proving that Wuhan has become an important habitat and breeding ground for Baer's pochards.

2021 年 10 月，沉湖湿地修复示范区和王家涉湖周边的水塘发现了 9 只青头潜鸭。这里曾是人工养殖鱼塘，实行退养还湖和生态修复后，栽植了很多水生植物。随着生态修复效果初见端倪，沉湖也逐渐成为青头潜鸭的重要栖息地。

In October 2021, nine Baer's pochards were found in the restoration demonstration area and the pond near Wangjiashe Lake in Chenhu Lake Wetland. The area, once an artificial fish farming pond, has aquatic plants grown since the return of the fish farming pond to the lake. The effort of ecological restoration seems to be paying off, and Chenhu Lake is becoming an import habitat for Baer's pochards.

罗纹鸭
Falcated Duck

罗纹鸭体型略小于绿头鸭及家鸭，雌性体色为暗褐色，杂有红褐色斑纹，两肋有类似贝壳状的罗纹；雄性罗纹鸭头部两侧为鲜亮的绿色，头顶为棕红色，身体两侧有大片罗纹，故而得名。

七壕，大群的罗纹鸭群飞舞　魏斌摄
A Great Number of Falcated Ducks Flying in Qihao　Photo by Wei Bin

The falcated duck (*Anas falcata*) is smaller than the mallard duck and the domestic duck. The female is dark brown with reddish-brown streak, and there are conchoidal patterns on both flanks. The male has an iridescent green head with a brownish-red patch on the top. It has falcated feathers on both sides of the body, and this morphological feature gives the duck its name.

2018 年冬，沉湖记录到 1800 只罗纹鸭，2019 年上升到 3300 只，到 2020 年，猛增至 22015 只， 2021 年，最高记录到 29249 只，出现爆发式增长。

A total of 1,800 falcated ducks were recorded on Chenhu Lake in the winter of 2018, and this figure rose to 3,300 in 2019. Falcated ducks in Chenhu Lake have seen explosive growth, with a sharp increase to 22,015 in 2020 and a record high of 29,249 in 2021.

4 平沙落“雁” Geese

雁类为鸭科大型游禽，善于飞行，是典型的迁徙水鸟。雁类以陆生植物的根、茎、叶以及一部分近岸的水生植物绿色部分为食。迁飞区内的雁类多繁殖于俄罗斯西伯利

灰雁和豆雁群　魏斌摄
Greylag Geese and Bean Geese　Photo by Wei Bin

亚地区或我国东北，其越冬地集中于中国的长江中下游地区。秋季退水后的洲滩生长着鲜嫩植物，为雁类提供了丰富的食物资源及栖息地。植物的空间分布、质量以及丰富度对植食性水鸟的分布、种群结构和个体行为都至关重要。在沉湖，共记录到鸿雁、豆雁、灰雁、白额雁、斑头雁、雪雁、白颊黑雁、帝雁等雁类。

白额雁群飞　魏斌摄
A Flock of Barnacle Geese Flying　Photo by Wei Bin

起飞的豆雁　魏斌摄
Bean Geese Taking Off　Photo by Wei Bin

短嘴豆雁　颜军摄
Short-Billed Bean Geese　Photo by Yan Jun

Geese are typical migratory waterfowl and large natatorial birds of the Anatidae family, which are good at flying. They feed on the roots, stems and leaves of terrestrial plants as well as the green parts of some waterfront aquatic plants. Geese on the flyway mostly breed in the Siberian region of Russia and northeastern China, and overwinter in the middle and lower reaches of the Yangtze River. In autumn, the bottomland with fresh vegetation after the

扫码观赏视频
Scan and watch

water recedes provides abundant food and ideal habitat for geese. The spatial distribution, quality, and abundance of vegetation are critical to the distribution, population structure, and individual behavior of phytophagous waterfowl. Goose species have been recorded on Chenhu Lake, including the swan goose, bean goose, greylag goose, white-fronted goose, bar-headed goose, snow goose, barnacle goose, and emperor goose.

豆雁是长江中下游湿地的常见种，数量多，分布十分广泛。根据长江中下游水鸟同步调查，沉湖是豆雁在湖北最重要的越冬地，支持了约78%~90% 的种群数量。每年有 8000~20000 只豆雁在沉湖湿地越冬，多分布于湖区退水后的草洲，也常在近岸浅水、泥滩地栖息。

The bean goose is a common species in the wetlands of the Yangtze River's middle and lower reaches, with a large population and wide distribution. According to a survey on waterfowl, Chenhu Lake is the most important wintering ground in Hubei, with a carrying capacity of 78%~90% of the bean geese population. Each year, there are 8,000 to 20,000 bean geese overwintering in the Chenhu Lake Wetland. Most of them inhabit the vegetated bottomland after the water recedes, and some also inhabit the shallow water and mudflats by the lake.

雁类越冬期的主要食物是薹草，有时也会在庄稼地、冬小麦地取食。薹草为莎草科薹草属多年生草本植物，为长江中下游地区常见种，在间歇性波动湖泊中发育较好。薹草属植物种子具有很强的休眠特性，主要以地下茎繁殖，湖区薹草一年有两轮生长轮。初春，薹草由地下茎先端或节上萌生植株，至 4~5 月份达生长旺盛期，6 月开花结果。夏季，湖泊进入汛期，薹草逐渐被湖水淹没而停止生长，转入休眠状态，地上部分逐渐腐烂。汛期过后，洲滩显露，薹草地下茎再次萌发新芽，9~10 月份进入秋季生长旺盛期，至冬季，地上部分枯萎，薹草再次进入休眠期。在薹草的整个生活史中，萌发的迟早、生长期的长短、生长状况和生物量等都对水位波动有一定的需求：首先，薹草不能在淹没区域萌发；其次，水位上涨速率若超过幼苗生长速率（每天 1.2 cm），则会对薹草生长产生不利影响；薹草成熟期需要一定淹没，淹没有利于薹草种子传播，淹没后薹草地上部分死亡，地下部分仍可存活，有利于来年萌发。淹没可使薹草获得竞争优势。

Geese in the wintering period mainly feed on sedge, and sometimes, they also obtain their food in cropland and winter wheat fields. Sedge, a perennial herb of genus Carex and family Cyperaceae, is common in the middle and lower reaches of the Yangtze River, with a wide distribution and better development in intermittently fluctuating lakes. Because of the strong dormancy of its seeds, sedge mainly propagates itself by means of the underground stem and completes two life cycles a year. Sedge sprouts from the apex or node of the underground stem in early spring, reach the peak stage of growth in April to May, and bloom and bear fruit in June. In summer, during the flood season, sedge is gradually inundated by water and turns into a dormant state, with the part above the ground rotting away. After the flood season, the bottomland gets exposed and the underground stem of sedge sprouts again, and the plant enters its peak growing period in autumn from September to October. It enters the dormant period again in winter when the above-ground part dies. During the whole life cycle of sedge, the time of germination, length of growth, growing conditions and biomass are all associated with water level fluctuations. Sedge does not germinate in inundated areas, and its growth will be adversely affected if the water level rises at a pace faster than the growth of its seedlings (1.2 centimeters per day). Moreover, its maturity needs certain inundation, which is conducive to seed dispersal. After it is inundated, the part above the ground dies and that under the ground survives. This is conducive to germination in the next year, and sedge can gain its competitiveness during the process.

某地偶尔出现的“迷鸟”，是指其繁殖地、越冬地及迁徙线路均不包含该地区，因“迷路”而出现的候鸟。台风等不良天气、某些个体导航能力的缺陷或幼鸟没有迁徙经验，都可能导致鸟类“迷路”。在繁忙的鸟类“加油站”或停歇地，落单或错认队伍的鸟类，也可能混入其他种群而被“带偏”。

Stragglers occasionally appearing in a certain place means that the bird's breeding ground, wintering ground or migration route does not concern the area where birds appear because they lost their way. Bad weather, such as typhoons, as well as deficiencies in the navigation ability of some individual birds or the inexperience of young birds in migration, may cause birds to “get lost”. In a busy bird stop, birds who are alone or enter the wrong team may be mixed with other bird species and then be misled.

沉湖湿地一直是迷鸟的福地。它们是沉湖的“不速之客”。罕见鸟种频频来沉湖“打卡”，高调“亮相”，让观鸟爱好者又惊又喜，连连称奇。

The Chenhu Lake Wetland has been a happy place for stragglers. Rare bird species have had frequent high-profile presence in Chenhu Lake as “uninvited guests”, making birdwatchers surprised and amazed.

卷羽鹈鹕

Dalmatian pelican

卷羽鹈鹕的体羽灰白色，飞羽羽尖黑褐色，头上冠羽呈卷曲状，因而得名。卷羽鹈鹕体长 1.6m~1.8m，体重可达 10kg，是体型最大的一种鹈鹕。卷羽鹈鹕分布于欧洲东南、中亚、中国东部地区。在其分布区内形成了西部、中部和东部 3 个彼此孤立的种群，其中繁殖于蒙古西部的东部种群形势最为危急。卷羽鹈鹕从繁殖地蒙古西部向我国东南部越冬地迁徙的路途中需要穿越近 1500 km 的干旱区域。

The Dalmatian pelican (*Pelecanus crispus*) has grayish-white feathers, dark brown wingtips, and curly feathers on its head, hence its name. With a body length of 1.6–1.8 meters and a weight of up to 10 kilograms, the Dalmatian pelican is the largest member of the pelican family. Dalmatian pelicans are distributed in Southeast Europe, Central Asia and

卷羽鹈鹕　颜军摄
Dalmatian Pelicans　Photo by Yan Jun

飞行中的卷羽鹈鹕 颜军摄
Dalmatian Pelicans Flying Photo by Yan Jun

eastern China. Three isolated species groups are formed in the western, central and eastern regions according to their distribution areas, among which the eastern species group that breed in western Mongolia faces the most critical situation. Dalmatian pelicans need to cross nearly 1,500 kilometers of arid areas to migrate from the breeding ground in western Mongolia to the wintering ground in southeastern China.

卷羽鹈鹕是国家一级重点保护野生动物。20世纪70至80年代，洪湖、沉湖、梁子湖、东湖都有一定数量的卷羽鹈鹕，此后已极罕见。2006年3月，沉湖记录到2只；2014年冬，又记录到2只；2015年11月，也记录到2只。

In China, it has been listed as a first-class national protected wild animal. From the 1970s to the 1980s, there were a certain number of Dalmatian pelicans in Honghu Lake, Chenhu Lake, Liangzihu Lake and East Lake. But since then, it was rarely seen. In March 2006, two Dalmatian pelicans were observed in Chenhu Lake. In the winter of 2014, another two were observed. There also were two Dalmatian pelicans observed in November 2015.

2021年10月，又有2只卷羽鹈鹕在北垸泵站现身。11月19日，增至5只，均在王家涉湖。12月，它们分开活动，王家涉湖2只，北垸泵站3只，并在沉湖度过了整个越冬季。卷羽鹈鹕以鱼类、甲壳类、软体动物、两栖动物等为食。觅食时以下颌橘黄色囊袋捞入大量水，滤去水后吞食其中的鱼虾。

In October 2021, two Dalmatian pelicans appeared at the Beiyuan Pumping Station. On November 19, 2021, the number increased to five, all found in Wangjiashe Lake. In December 2021, they moved separately, with two in Wangjiashe Lake and three at the Beiyuan Pumping Station. They spent the whole winter in Chenhu Lake. Dalmatian pelicans feed on fish, crustaceans, mollusks, amphibians, etc. When foraging, the Dalmatian pelican will scoop up a large amount of water with its orange pouch on its lower jaw, and swallow the fish and shrimps after the water is filtered out.

大红鹳

Greater flamingo

2021 年 11 月 4 日，保护区工作人员通过“智慧湿地”系统发现了一种不太一样的鸟。经过鉴定，“来客”是武汉十分罕见的迷鸟：大红鹳。此前，大红鹳在武汉的唯一一次记录，是 2015 年出现在府河湿地。

On November 4, 2021, the staff of the nature reserve discovered a different kind of bird through the “Smart Wetland” system. After identification, “the guest” is a rare straggler in Wuhan: Greater flamingo (*Phoenicopterus roseus*). Previously, the only record of Greater flamingo in Wuhan was in 2015 when it appeared in the Fuhe Wetland.

大红鹳分布于西非、撒哈拉沙漠以南、地中海以及亚洲的西南部和南部地区。它们的迁徙路线不经过我国，所以只有在它迷路时，我们才有机会“一睹芳容”。它时而在湖心单腿小憩，时而在浅水区漫步觅食，起飞时，如同低空一抹粉红的云朵。这只大红鹳在沉湖度过了整个越冬季。

Greater flamingos are distributed in West Africa, sub-Saharan Africa, the Mediterranean Sea, and southwest and south Asia. Their migration route does not pass through China, so only when they get lost can we have a chance to see their beauty. Sometimes they take a nap with a single leg standing in the middle of the lake, and sometimes they stroll in the shallow water for food. They look like pink clouds when flying. This greater flamingo above-mentioned spent the whole winter in Chenhu Lake.

大红鹳食性相对广泛，主要包括甲壳类、软体动物、昆虫幼虫等动物性食物以及植物种子、藻类和腐败的树叶等植物性食物，偶尔还会取

大红鹳　魏斌摄
A Greater Flamingo　Photo by Wei Bin

食昆虫的成虫、螃蟹和小型鱼类。

Greater flamingos have a wide range of feeding habit, mainly feeding on animal foods such as crustaceans, mollusks, and insect larvae, as well as plant foods such as plant seeds, algae and rotten leaves, and occasionally feeding on adult insects, crabs and small fish.

白颊黑雁

Barnacle goose

白颊黑雁面部有白斑、背部灰色、颈部和胸部黑色，繁殖于北极沿海地区的悬崖上，耐严寒，喜栖于海湾、海港及河口等地，主要以青草或水生植物的嫩芽、叶、茎等为食。在 2009 年、2018 年、2019 年和 2021 年，沉湖均记录到白颊黑雁。

The barnacle goose (*Branta leucopsis*) has a white face, gray back, and black neck and chest. They breed on the cliffs in Arctic coastal areas, resistant to severe cold, and like to live in bays, harbors and estuaries. They mainly feed on the buds, leaves and stems of grass or aquatic plants. Barnacle geese were recorded in Chenhu Lake in 2009, 2018, 2019 and 2021 respectively.

难得一见的白颊黑雁，和其他雁类混群　魏斌摄
Barnacle Geese Mingled with Other Geese　Photo by Wei Bin

第四章

Chapter IV

和合向未来

Harmony and Cooperation for a Better Future

2000 年 11 月，由国家林业局牵头，17 个部门共同参与编制的《中国湿地保护行动计划》正式颁布，明确了湿地保护的指导思想和战略任务，这是我国湿地保护与可持续利用的一个纲领性文件。2003 年 8 月，国家林业局会同国家发改委等单位完成了《全国湿地保护工程规划（2002—2030）》，提出了湿地保护的长远目标。2004 年，国务院办公厅发出了《关于加强湿地保护管理的通知》，提出对自然湿地进行抢救性保护。2005 年，国务院批准了《全国湿地保护工程实施规划（2005—2010）》，以工程措施对重要退化湿地实施抢救性保护。中国自 1992 年加入《湿地公约》以来，指定了国际重要湿地 64 处，建立了湿地类型自然保护区 600 多处、国家湿地公园 1600 余处，初步形成了一个覆盖全国大部分重要湿地的保护网络体系，重要的天然湿地和一大批濒危重点保护物种得到了较为有效地保护。

武汉市沉湖湿地的保护历程，正是我国湿地保护与国际公约履约工作的一个小小缩影。“和”是全民珍爱湿地，人与自然和谐共生；“合”是兼容并蓄，携手合作，充分运用科技力量，自信创新，迈向未来。

In November 2000, the *China National Wetlands Conservation Action Plan*, jointly prepared by 17 departments under the lead of the State Forestry Administration (now replaced by the National Forestry and Grassland Administration), was officially promulgated. As a programme for conservation and sustainable utilization of Chinese wetlands, it explained the guideline and strategic tasks for the wetland conservation. In August 2003, the State Forestry Administration, together with the National Development and Reform Commission and other agencies, completed the *National Wetland Conservation Project Plan (2002 –2030)*, proposing long-term goals for wetland conservation. In 2004, the General Office of the State Council issued the *Notice on Strengthening the Management of Wetland Conservation*, which proposed the remediation and protection of natural wetlands. In 2005, the State Council approved the *National Wetland Conservation Project Implementation Plan (2005 –2010)*, implementing the remediation and protection of important degraded wetlands via engineering measures. Since its accession to the *Ramsar Convention on Wetlands* in 1992, China has designated 64 wetlands of international importance, established over 600 wetland nature reserves and more than 1,600 national wetland parks, and created the outlines of a protection network system covering most of the country's important wetlands, thus putting in place effective protection for important natural wetlands and a large number of endangered key protected species.

The conservation of Wuhan's Chenhu Lake Wetland represents the epitome of wetland conservation in China, and of Chinese implementation of the related international conventions. "Harmony" implies the cherishing of wetlands by all people, and the harmonious coexistence of humanity and nature; "cooperation" represents compatibility and collaboration, and full use of the power of science and technology to confidently progress towards a better future through innovation.

1 保护条例 一湖一策
Protective Regulations for Lakes

1999 年，武汉市出台了《武汉市保护城市自然山体湖泊办法》，正式拉开了湖泊保护的序幕；2002 年，颁布实施了《武汉市湖泊保护条例》，这是我国第一部对湖泊进行全面、综合性管理的地方性法规；同年发布了《武汉市湖泊保护名录》，对市域 166 个湖泊建档管理；2005 年发布《武汉市湖泊保护条例实施细则》；2015 年，对条例进行了修正，对湖泊保护提出了更高的要求。

In 1999, Wuhan City issued the *Measures for Protecting Urban Natural Mountains and Lakes in Wuhan*, marking the official inception of lake protection work. In 2002, the promulgation and implementation of the *Wuhan Lake Protection Regulations* represented China's first comprehensive local lake management regulations. In the same year, the *Wuhan Lake Protection List* was released, registering the city's 166 lakes for management. In 2005, the *Detailed Rules for the Implementation of Wuhan Lake Protection Regulations* was issued, which was amended in 2015 to meet higher lake protection requirements.

流域治理的跨界特性非常典型，通过水这条纽带，与水有关的不同机构产生了无法分离的联系。而“河湖长制”则是中国流域治理跨部门协作的一个典型案例。2012 年，武汉市开始推行“湖长制”，为湖泊划定“三线一路”，成为全国湖泊管理保护的先行者。2015 年，首次将全市 166 个湖泊的湖岸线分段细化，对湖岸线的管理精确分割

鸟类调查经常需要徒步很远距离 供图：武汉市观鸟协会
Bird Surveys Often Require Walking Long Distances
Photo Credit: Wuhan Bird Watching Society

到每一米，将 166 个湖泊以行政区域管辖的现状为基础，结合湖泊蓝线，划定了 93 位 “官方湖长”湖泊岸线的管理责任范围。2016 年至 2017 年，国家先后印发《关于全面推行河长制的意见》、《关于在湖泊实施湖长制的指导意见》，部署全面推行河长制、湖长制，并要求实行一河一策、一湖一策。2017 年 1 月，湖北省印发《关于全面推行河湖长制的实施意见》，明确要求“立足不同地区不同河湖实际，统筹上下游、左右岸，做到一河湖一档、一河湖一策，解决好河湖管理保护中的突出问题”。在此基础上，武汉市构建了市、区、街道（乡镇）三级河湖长制责任体系，贯彻 “一湖一策”，以问题为导向，立足河湖实际，综合考虑湖泊面积、人口数量、湖泊自净力等因素，统筹上下游、左右岸，着力解决河湖管理保护中的突出问题，通过大力保护和整治，使湖泊水体的水质得到提升。

The river basin governance concerns the cooperation between different departments, with several water-related agencies inextricably linked via the water. The “river/lake chief system”, now typical of inter-departmental collaboration, is designed to promote the river basin management in China. In 2012, Wuhan designed the “three lines and one road” policy for lakes to implement its “lake chief system”, which made it a pioneer of lake management and protection in China. In 2015, the shoreline of the city’s 166 lakes for the first time was divided into small sections (accurate to the meter), with the management responsibility of each section allocated to different “chiefs”. Based on the prevailing administrative jurisdiction over the 166 lakes, in combination with the “blue lines” (geographical borders associated with water bodies), the scope of managerial responsibility of 93 “official lake chiefs” was defined. During 2016 and 2017, China issued the *Opinions on the Full Implementation of the River Chief System* and *Guiding Opinions on the Implementation of the Lake Chief System for Lakes*, made arrangements to fully implement the river and lake chief systems, and required the implementation of “one river, one policy” and “one lake, one policy”. In January 2017, Hubei Province issued the *Opinions on the Full Implementation of the River and Lake Chief System*, clearly requiring that “the major problems of river and lake management and protection be solved based on the actual situation of various rivers and lakes in different regions, coordinating the management of upstream and downstream sections, left and right banks, handling different rivers and lakes with different policies". On this basis, a three-tier responsibility system of river and lake chief systems has been established, with the city, district and sub-district (township) levels in Wuhan to implement the idea of "one lake, one policy". Targeting the problems, based on the reality of the rivers and lakes, Wuhan takes into account factors such as lake areas, self-purification capacity and populations, coordinates the management of upstream and downstream sections, and left and right banks, strives to solve the major problems of river and lake management and protection, and ensures that the water quality of lakes is improved via vigorous protection and remediation.

沉湖不断推进湖长制，将防治水污染、改善水环境、修复水生态作为主要任务，对入湖排口、港渠的水质进行监测，注重对流域范围内各类污染源和排水管网存在的问题进行清理整治，提升污水收集、处理效能，并对湖区水体自身污染进行清理整治。通过对污染防治攻坚，使得水环境持续改善。

The "lake chief" system has been continuously promoted at Chenhu Lake. Taking as its main tasks the prevention and control of water pollution, the improvement of water environment and restoration of water ecology, it also encompasses the monitoring of water quality at inlets and lake canals, tackling problems of various pollution sources and drainage pipes in the basin. Through promoting sewage collection and treatment, and reducing water pollution, the water environments have improved continuously.

2 为湖松绑　还绿于民
Return the Clear Lake to the People

1998 年的长江洪水使人们重新思考，要尊重自然规律，适当向江湖"让步"。在中央提出的"三十二字方针"，即"封山育林、退耕还林，平垸行洪、退田还湖，加固堤防、疏浚河湖，以工代赈、移民建镇"指导下，2020 年，为扭转长江生态环境恶化的趋势、保护长江生物多样性，长江流域重点水域开始实行十年禁渔计划。第一阶段自 2020 年 1 月 1 日零时起，长江上游珍稀特有鱼类国家级自然保护区等 332 个自然保护区和水产种质资源保护区全面禁止生产性捕捞。第二阶段自 2021 年 1 月 1 日零时起，在长江干流和重要支流、大型通江湖泊除水生生物自然保护区和水产种质资源保护区以外的天然水域，实行暂定为期十年的常年禁捕。武汉市实施禁渔以后，次年夏天，人们在武汉江段多次看到难得一见的江豚出没。

After the flood of the Yangtze River in 1998, a rethink is required to respect the laws of nature and make concessions to the rivers and lakes appropriately. Under the guidance of the “32-character policy” (32 Chinese characters are used in the original), that is, “close mountains and plant trees, convert cultivated lands to forests, make river-bed flat and control flooding, convert paddy fields to lakes, make dikes higher and stronger, make flows smooth in rivers and lakes, relieve people in disaster areas by giving them employment instead of

长江江豚　李锋摄
The Yangtze Finless Porpoise　Photo by Li Feng

outright grants, and build new towns for migrants" put forward by the central government, in 2020, a 10-year fishing ban plan started in key waters of the Yangtze River Basin in order to reverse the deterioration trend of the ecological environment and protect the biodiversity of the Yangtze. In the first stage, from 00:00 on January 1, 2020, 332 nature reserves and aquatic germplasm resources conservation areas, including the National Nature Reserve for the Rare and Endemic Fishes in the upper reaches of the Yangtze River, have completely banned productive fishing. In the second stage, from 00:00 on January 1, 2021, a temporary 10-year perennial fishing ban has been implemented in the natural waters of the main stream and important tributaries of the Yangtze River and large lakes that connects the Yangtze River, except for aquatic nature reserves and aquatic germplasm resources conservation areas. After the fishing ban was implemented in Wuhan, the Yangtze finless porpoises that had been rarely seen was found many times in the Wuhan section of the Yangtze River the following summer.

近年来，为恢复和优化湿地生态系统功能，改善鸟类栖息环境，沉湖及周边湿地开展退耕还湿试点约 9000 亩，投入资金 900 万元。并以“壮士断腕”的决心，强力推进退养还湖，将湿地自然保护区核心区和缓冲区内历史遗留的鱼塘、藕塘等养殖、种植项目全部取缔，拆除围网、网箱和拦网。退养面积共计约 7.8 万亩，包括水域 4.3 万亩，水产养殖塘 3.5 万亩，累计投入 7000 余万元，拆除看护房等人工设施 410 处约 3.2

一人观测，一人记录
供图：武汉市观鸟协会
Observing and Recording
Photo Credit: Wuhan Bird Watching Society

万平方米。湖区范围内的土地（水域）均流转或托管至沉湖湿地局统一管理，每年流转费 694 万元纳入区级财政预算。

In recent years, in order to restore and optimize the function of the wetland ecosystem and improve the habitat of birds, a pilot project of converting about 9,000 *mu* of farmland into the wetland has been carried out in Chenhu Lake and its surrounding areas, with an investment of RMB 9 millions. With great determination, Wuhan has vigorously pushed forward the conversion of aquaculture areas to lakes, banned all the aquaculture activities of fish ponds and lotus ponds left over from history in the core area of the wetland nature reserves and the buffer areas, and removed the purse seines, net cages and block nets. A total of about 78,000 *mu* of aquaculture areas have been converted, including 43,000 *mu* of water areas and 35,000 *mu* of aquaculture ponds. A total of more than RMB 70 millions has been invested, and 410 artificial facilities (such as aquaculture tending rooms) , covering an area of about 32,000 square meters, have been removed. The land (water area) within the basin has been transferred or entrusted to the Chenhu Lake Wetland Bureau for unified management, and the annual transfer fee of RMB 6.94 millions has been included in the district-level financial budget.

拆除人工设施　供图：沉湖湿地省级自然保护区管理局
The Removal of Artificial Facilities　Photo Credit: Chenhu Lake Wetland Provincial Nature Reserve Administration

消泗乡渔樵村位于张家大湖畔，村支书陈为炳对沉湖感情深厚而又复杂：“50 年前的儿时记忆是，捧起沉湖的水就喝，那时村里没自来水，生活用水全靠它。20 世纪 90 年代，鱼塘藕塘渐多，沉湖被各种围网围埂围困。最近几年为它松绑，如今终

拆除人工设施之前　供图：沉湖湿地省级自然保护区管理局
Before the Removal of Artificial Facilities　Photo Credit: Chenhu Lake Wetland Provincial Nature Reserve Administration

拆除人工设施之后　供图：沉湖湿地省级自然保护区管理局
After the Removal of Artificial Facilities　Photo Credit: Chenhu Lake Wetland Provincial Nature Reserve Administration

武汉市观鸟协会工作人员在沉湖
供图：武汉市观鸟协会
Staff of Wuhan Bird Watching Society in Chenhu Lake　Photo Credit: Wuhan Bird Watching Society

于还给天地之间了。”这份“还”来得又是何等不易。

Yuqiao Village of Xiaosi Township is situated by Zhangjia Lake. The village secretary Chen Weibing has a deep and complicated affection for Chenhu Lake: “I remembered that 50 years ago when I was a child, I drank the water from Chenhu Lake. There was no tap water in the village at that time, and we got all our daily drinking water from the lake. Since the 1990s, there were more and more fish ponds and lotus ponds, and the lake was surrounded by various purse seines. The purse seines have been removed in recent years, and now we have finally returned the lake to nature.” It really took us great efforts to do this.

张家大湖是沉湖的一个重要子湖。往年冬天，湖上一条条运藕船接力出湖，而今冬湖面则随处可见红嘴鸥、白骨顶等鸟类自由栖息。2014 年 12 月，全湖水鸟 37 种，14717 只；2018 年 1 月，水鸟 36 种，16570 只；2021 年 1 月，48 种，77135 只。3 年间，从 1 万多只到近 8 万只，武汉市观鸟协会这组观测沉湖候鸟的数据是生态保护和修复成果的例证之一。冯江自 2008 年来沉湖湿地保护区工作，看到以往被鱼塘藕塘分隔的沉湖终于破埂松绑了，他高兴地说：“给它自由，没人为干扰，恢复效果非常好。候鸟种类数量恢复性提升，很有成就感！当然，担子也更重。”

In Zhangjia Lake, an important sub-lake of Chenhu Lake, we could often see a number of boats transporting lotus roots in the winter of previous years. However, this winter, birds like black-headed gulls and Eurasian coots are found here, living freely. In December 2014, there were 14,717 waterbirds under 37 species in the lake. In January 2018, there were 16,570 waterbirds under 36 species. In January 2021, there were 77,135 waterbirds under

48 species. During the past three years, the number of waterbirds has increased from more than 10,000 to nearly 80,000. This group of data of migratory birds observed by the Wuhan Bird Watching Society is an illustration of the environmental protection and restoration achievements. Feng Jiang has been working in the Chenhu Lake Wetland Nature Reserve since 2008. Seeing that Chenhu Lake, once separated by fish ponds and lotus ponds, finally has got the purse seines removed, he says happily, “We give the lake enough freedom without human interference. The restoration effect is quite good. Since the number of migratory bird species has been increasing, I feel a great sense of accomplishment! Of course, there is still a lot more to do. ”

退养还湖是沉湖回归自然的根本，剔除以往有经济效益的鱼塘和藕塘，将湿地交还给大自然，有“舍”才有“得”。挣脱渔网和围埂的束缚，而今复返自然的沉湖，

沉湖　魏斌摄
Chenhu Lake　Photo by Wei Bin

灵动舒展野性美，吸引候鸟“大部队”入驻，众多珍禽流连忘返。

Returning aquaculture areas to the lake is the foundation of returning the lake to nature. We should give up the fish ponds and lotus ponds that used to bring economic benefits, and return the wetlands to nature. You have to give to take. Breaking free from the shackles of fishing nets and fences, Chenhu Lake has attracted a large number of migratory birds with its natural charm and wild beauty. There also are many rare birds lingering here.

3 自然恢复　生境管理
Nature Restoration and Habitat Management

“水，是湿地的灵魂。”保护区管理局局长方瑛介绍，“湿地生境恢复首先要破围堤，实施水系连通，并疏浚河道。流水不腐，沉湖的水活了，这是基础。”

“Water is the soul of the wetland,” says Fang Ying, Director of the Management Bureau of the Chenhu Lake Wetland Nature Reserve. “To restore the bird habitat in the wetland, we need to remove the fences, connect the water systems, and dredge the rivers. Running water never becomes putrid, and it is important to keep alive the water in the lake .”

对淤塞的 24 公里水道进行疏浚，并通过挖沟渠、移植、封育等办法积极恢复湿地生态功能。坚持原生态、微改造、少干预的原则，自然恢复为主，人工修复为辅。

We have dredged the blocked 24-kilometer watercourse, and restored the ecological function of the wetland by digging ditches, transplanting trees and setting up enclosures. We adhere to the principle of original ecology, micro-renovation, and less intervention, with the natural restoration as the mainstay and artificial restoration as the supplement.

大型水生植物是水体生态系统的初级生产者之一，对维持淡水生态系统的结构和功能有至关重要的作用。保护区选择该区域原分布的植物种类如芦苇、荻、菰、菱、薹草等作为植被恢复物种，按照水位高低依次栽种，形成带状分布的混生群落。根据曲口外滩、七壕外滩、三汊河、洪道等重要的鸟类栖息地的实际情况，通过改造微地形来改善水文环境，因地制宜地补植补种薹草、芦苇、荻等植物，为众多鸟类，特别是雁鸭类等提供了丰富的食物来源和庇护所。在观荷长廊，保护与恢复的主要植物是野莲，这里是夏季黑水鸡、黄斑苇鳽等水鸟良好的觅食及繁殖场所。目前，湖区已完成退化湿地恢复面积 1400 公顷；湿地植被恢复面积 110 公顷；鸟类食物源补充面积 25 公顷。

Macrophyte is one of the primary producers of water ecosystems, which plays a vital role in maintaining the structure and function of the freshwater ecosystems. The plant species originally distributed in this area, such as *Phragmites communis*, *Triarrhena sacchariflora*, *Zizania latifolia*, *Trapa bispinosa* and *Carex argyi*, are selected as vegetation restoration species, which are planted in sequence according to the water level to form a mixed

一群白琵鹭飞过沉湖宾馆，生态修复后，这些建筑已经拆除　魏斌摄
Eurasian Spoonbills Flying over the Chenhu Lake Hotel, Which Has Been Removed after the Ecological Restoration　Photo by Wei Bin

ecological community featuring zonal distribution. According to the actual situation of important bird habitats such as Qukou Bund, Qihao Bund, Sanchahe Lake and the Spillway, the hydrological environment is improved by modifying the micro-topography, and plants such as *Carex argyi*, *Phragmites communis*, and *Triarrhena sacchariflora* are replanted according to local conditions, providing abundant food sources and shelters for many birds, especially geese and ducks. The main plant protected and restored in the Lotus Viewing Corridor is the wild lotus, which provides a good foraging and breeding place for waterbirds such as common moorhens and yellow bitterns in summer. At present, 1,400 hectares of degraded wetland have been restored; the restoration area of wetland vegetation has reached 110 hectares; the replenishment area of bird food sources has reached 25 hectares.

同时，借破堤顺势垒起的几座生境岛，在冬季水位下降时露出大片滩涂，可供游禽涉禽自由活动，给动物营造出了相对独立、低干扰的隐蔽空间。环境变好了，野猪和蛇等动物在丰水期也有了归宿。

Meanwhile, several habitat islands have been built after the fences are removed. The water level drops in winter to expose a large mudflat for the free movement of migratory birds and wading birds. We have created a relatively independent, low-interference hidden space for animals. Since the environment has become better, animals like wild boars and snakes have also found their homes in wet seasons.

自然的修复看似轻描淡写，没有人造景观，却是事半功倍。野性沉湖逐步形成，个中艰辛与坎坷，也是格局与奉献的诠释。

Natural restoration seems to be easy, and there is no artificial landscape involved, but it yields twice the result with half the effort. The wild Chenhu Lake is gradually taking shape, and the hardships and frustrations we encountered in the process are manifestations of our dedication and contribution.

外来物种治理

Handling Exotic Species

加拿大一枝黄花是原产北美洲的一种外来入侵物种，它繁殖力极强，

加拿大一枝黄花
供图：武汉市园林和林业局
Canadian Goldenrod
Photo Credit: Wuhan Municipal Bureau of Gardening and Forestry

传播速度极快，竞争力极强，很容易挤占本土植物的生存空间，被称为“生态杀手”。沉湖湿地组成专业队伍开展专项调查、除治工作，以严格的技术标准，实施人工和机器拔除，在花败结果前彻底铲除，并集中深埋或焚烧，防止其再生和蔓延。

Canadian goldenrod (*Solidago canadensis*) is an exotic invasive species native to North America with strong fecundity, rapid spreading speed and strong competitiveness, and it is easy for it to occupy the living space of indigenous plants. It is called an “ecological killer”. A professional team has been set up to carry out the special investigation and extermination work in Chenhu Lake. Manual and mechanical uprooting has been carried out according to strict technical standards. The Canadian goldenrods were completely eradicated before fruit-bearing, and buried or burned in a centralized manner to prevent their regeneration and spread.

4 生态补偿 请鸟吃饭
Ecological Compensation for Crops Eaten by Birds

武汉市积极探索湿地自然保护区生态补偿，2013 年 10 月，出台了《武汉市湿地自然保护区生态补偿暂行办法》。为规范资金使用，次年又出台了《武汉市湿地自然保护区生态补偿资金管理办法》。2014 年，我国正式启动湿地生态效益补偿试点工作，沉湖湿地自然保护区也在同年开始试点。

Wuhan has actively explored ecological compensation of wetland nature reserves. In October 2013, the *Interim Measures for Ecological Compensation of Wuhan Wetland Nature Reserves* was issued. In order to standardize the use of funds, the *Measures for the Management of Ecological Compensation Funds in Wuhan Wetland Nature Reserves* was issued the following year. In 2014, China officially launched the pilot project of ecological benefit compensation for wetlands, and the Chenhu Lake Wetland Nature Reserve also started the pilot project in the same year.

保护区内的农户是沉湖开展生态补偿工作的重要对象。他们的直接经济损失，主要是由于野生动物捕食、破坏导致农业产值受到影响。冬候鸟中，豆雁等植食性鸟类喜食油菜、小麦等作物；普通鸬鹚、鹭类及其他肉食性和杂食性鸟类会对渔业养殖造成损失；雉鸡、野猪等野生动物还会对玉米、西瓜等农作物造成破坏。间接经济损失是土地发展权受限，保护区建立限制区域土地发展权，农户失去改变土地用途、增加收益的权利。

白琵鹭群飞　颜军摄
Flocks of Eurasian Spoonbills Flying　Photo by Yan Jun

Farmers in the nature reserve are important objects of ecological compensation in Chenhu Lake. Their direct economic loss is mainly due to the predation and destruction caused by wild animals, which affects the agricultural output value. Among winter migrants, herbivorous birds such as bean geese like to eat rape seeds and wheat. Common cormorants, herons and other carnivorous and omnivorous birds cause losses to fishery culture. Wild animals such as pheasants and wild boars destroy crops, e.g. corn and watermelon. The indirect economic loss is caused by the restriction on the land development right of the farmers due to the establishment of the nature reserve, because the farmers can not anymore change their land use to increase their income.

合理有效的补偿方式和科学的生态补偿标准是维护湿地周边农户合法权益、激励农户保护湿地资源、维护国家生态安全的现实要求和重要保障。自 2014 年起，武汉市、蔡甸区政府每年出资 473 万元实施生态补偿试点工作。2017 年提高到 642.9 万元，用于激励周边农户保护湿地和野生动物，引导村民改变生产生活方式，把绿水青山变成

小天鹅扇翅 魏斌摄
The Tundra Swan Flapping Its Wings Photo by Wei Bin

金山银山。随着环境质量的提高，沉湖湿地周边的经济发展大幅跃升，居民生活质量也稳步提高。

Reasonable and effective compensation methods and scientific ecological compensation standards are realistic requirements and important guarantees for safeguarding the legitimate rights and interests of farmers around the wetlands, encouraging farmers to protect wetland resources and maintaining national ecological security. Since 2014, Wuhan Municipal Government and the Government of Caidian District have invested RMB 4.73 millions each year to implement the pilot project of ecological compensation. In 2017, the investment was raised to RMB 6.429 millions, which was used to encourage surrounding farmers to protect wetlands and wild animals, help villagers to change their production method and lifestyle, and turn green mountains and clear water into invaluable assets. As the environment quality improves, the economy around the Chenhu Lake Wetland has greatly developed, and the life quality of residents has also steadily improved.

近苇荡，远鸬鹚树，还有大群水鸟　魏斌摄
Reed Marshes, Cormorants in Trees and Other Water Birds　Photo by Wei Bin

5 万众之力　全民共举
Joint Efforts of the Chinese People

保护网络
Protection of Networks

2007 年，由沿长江六省市 (上海市、湖北省、湖南省、江西省、安徽省、江苏省) 林业厅局、世界自然基金会共同发起并组建成立了长江中下游湿地保护网络。这是一个由管理机构、研究单位、社会团体和公众广泛参与的流域性战略合作平台。平台旨在通过信息共享、经验交流、能力建设和实地示范等活动，推动长江流域 2 万多平方公里湿地得到有效保护与管理，抢救性保护现有湿地，遏制长江湿地功能进一步退化，促进长江中下游生态文明和生态体系建设。从生态流域角度建立协调机制，保护湿地生态，这是国内首创。沉湖湿地是长江中下游湿地保护网络首批 22 个重要成员之一。

In 2007, the Yangtze River's middle and lower reaches wetland conservation network was jointly initiated and established by the forestry administrations of five provinces (Hubei, Hunan, Jiangxi, Anhui and Jiangsu) and Shanghai City along the Yangtze River, as well as the World Wide Fund for Nature. As a basin-wide strategic cooperation platform with broad

志愿者在记录鸟况
供图：武汉市观鸟协会
Volunteers Recording about Birds　Photo Credit: Wuhan Bird Watching Society

participation including management institutions, research institutes, social organizations and the public, it aims to promote the effective protection and management of over 20,000 square kilometers of the Yangtze River Basin wetlands via activities such as information sharing, experience exchange, capacity building and field demonstration, and to protect the existing wetlands, thereby curbing further degradation of their functioning, and promoting ecological civilization and ecosystem construction in the Yangtze River's middle and lower reaches. It was the first initiative in China to establish a coordination mechanism for the protection of wetland ecology, and the Chenhu Lake Wetland was one of the initial 22 important network members in the middle and lower reaches of the Yangtze.

民间湖长
Citizen Lake Chiefs

不断拓宽社会公众参与渠道，积极营造“共建共治共享”美丽河湖的良好氛围，2012 年，武汉市开始向社会公开征集热心于河湖保护公益事业的志愿者和相关公益活动的志愿团体担任“民间湖长”。目前，武汉市三级“民间河湖长”已达到 1145 名。

To constantly broaden channels of public participation, and create a positive awareness of beautiful rivers and lakes constructed, managed, and shared by all, Wuhan in 2012 started public recruitment of volunteers keen on river and lake protection, and volunteer groups involved in the relevant public welfare activities. These volunteers and volunteer groups were to serve as "citizen lake chiefs". At present, Wuhan has boasted 1,145 "citizen lake chiefs" at the city, district and sub-district (township) levels.

社会力量
Social Impetus

自 20 世纪 80 年代以来，中国民间观鸟、护鸟活动蓬勃发展，鸟类摄影爱好者的队伍快速壮大。武汉鸟类观测记录始于 19 世纪 60 年代。2007 年，武汉野生动植物保护协会观鸟分会成立，2017 年，注册为独立的社团组织——武汉市观鸟协会。从 2016 年起，武汉市观鸟协会定期发

冬天的鸟类调查 供图：武汉市观鸟协会
Birds Survey in Winter Photo Credit: Wuhan Bird Watching Society

布武汉重点区域鸟类监测月报和年报，供市民了解“鸟情”，并在原有的基础上，开始按照规范程序，认证武汉鸟类的观测记录。武汉市观鸟协会与沉湖保护区管理局合作，共同监测湿地鸟类。此外，武汉还有多个自然环境保护组织和20多万自然保护和科普宣教志愿者，其中爱河护湖志愿者队伍规模突破5000人，这都是湿地保护的中坚力量。其中，武汉“爱我百湖志愿者协会”公益行动多年来不断开展巡查河湖行动、持续关注河湖问题、努力提高河湖生态，充分发挥民间环保力量与官方管理者间的桥梁和纽带作用，开展了武汉湖泊第三方调查、沉湖湿地水鸟及其栖息地保护项目研究等科研项目和活动，2017年被中宣部、中央文明办评为全国100个“最佳志愿服务组织”之一，是武汉市首个获此项殊荣的环保类志愿者服务组织。

Since the 1980s, bird watching and protection activities amongst the Chinese population have developed vigorously, and bird photographers have grown rapidly in number. There were bird observation records in Wuhan in the 1860s. In 2007, the Wuhan Wildlife Conservation Association's Bird Watching Branch was established. In 2017, this branch was registered as an independent community organization— the Wuhan Bird Watching Society. Since 2016, the Society has released monthly and annual bird monitoring reports on the key areas of Wuhan to inform the public

两人观测，一人记录 供图：武汉市观鸟协会
Observing and Recording Photo Credit: Wuhan Bird Watching Society

of "the birds' situation", and, based on the past practice, to begin to certify the records of bird observations in Wuhan according to standard procedures. The Wuhan Bird Watching Society is now working with the Chenhu Lake Wetland Nature Reserve Management Bureau to monitor the wetland birds. There are also several natural environment protection organizations in Wuhan, and over 200, 000 volunteers involved in nature protection and popular science education, of whom there are over 5, 000 involved in river and lake protection. They serve as the backbone of the wetland conservation. The Wuhan Lake-loving Volunteers Society, for example, has been conducting inspections of rivers and lakes for years, continually paying attention to the problems of rivers and lakes, striving to improve their ecological environment, playing its role as a bridge and link between non-government environmental protection forces and official administrators, and carrying out scientific research projects and activities, including the third-party investigation into Wuhan's lakes, the research on waterfowl and the Chenhu Lake Wetland habitat protection project. In 2017, the Society was rated as one of China's 100 "Best Volunteer Service Organizations" by the CPC Central Committee's Publicity Department and the Office of Spiritual Civilization Development Steering Committee, becoming the first organization in Wuhan to be awarded this honor.

6 智慧湿地　科技助力
Smart Wetlands Powred by Science and Technology

监测是生态保护工作的重要组成部分。沉湖自然保护区一共有 3 个管护站 36 名人员负责巡湖，一年 365 天 24 小时值班。每天早、中、晚巡湖 3 次，巡湖里程约 100 公里。以往巡湖靠摩托车和徒步，目前巡护装备齐全，有皮卡巡护车、摩托车和快艇等。近年来，沉湖又加强了湿地监测相关的基础设施，先后完成罗汉生态监测站、沉湖渔场生态监测站、王家涉生态监测站、七壕保护站、马弓林场保护站、曲口保护站、黄家合子管理站、鸟类救护中心、湿地实时监控系统、观鸟长廊、巡护道路等基建工程。还进一步与科研机构和社会团体合作，开展自测、委托监测和联合监测；

智慧巡护工作
供图：缤纷自然
Smart Patrol Work
Photo Credit:
Colorful Nature

并加强与水务、环保、气象等部门的联动，为湿地综合整治提供数据支撑，为突发环境事件提供预警。

Monitoring is an important element of ecological protection, and the Chenhu Lake Wetland Nature Reserve's three management and protection stations have a staff of 36 on duty 24 hours a day all year round. Three daily patrols are performed (morning, noon, and evening), each covering around 100 kilometers. The personnel in the past patrolled by motorcycle and on foot, but now pickup trucks and speedboats are available. In recent years, wetland monitoring infrastructure availability has improved, thanks to the completion of infrastructure construction projects including ecological monitoring stations at Luohan, Chenhu Fishing Ground and Wangjiashe, protection stations at Qihao, Magong Forest Farm and Qukou, along with Huangjiahezi Management Station, the Bird Rescue Center, the Wetland Real-time Monitoring System, the Bird Watching Corridor and the Patrol Roads. Further cooperation with scientific research institutions and social organizations allows the performance of self-monitoring, delegated monitoring and joint monitoring, while strengthened collaboration with the departments concerning water affairs, environmental protection, and meteorology, etc. has provided data support for comprehensive wetland improvement and early environmental emergency warning.

智慧城市是运用物联网、云计算、大数据、空间地理信息集成等新一代信息技术，促进城市规划、建设、管理和服务智慧化的新理念和新模式。2012年，《关于国家智慧城市试点暂行管理办法》的出台拉开了我国智慧城市建设的序幕。

The “smart city” is a new concept/model that adopts advanced information technologies, such as the Internet of Things (IoT), cloud computing, big data, geographic information integration, etc. to promote the city planning, construction, management and services. In 2021, the *Interim Management Measures for National Smart City Pilots* was issued which initiated China’s construction of smart cities.

2015 年，武汉市成为国家首批“智慧城市”试点，开启了大数据环境下的定量规划与治理决策智能化探索。通过多源大数据的融合与空间模型构建，模拟复杂城市系统、感知城市体征、监测城市活动、预演城市未来。

In 2015, Wuhan was selected as one of the first batch of pilot Chinese “smart cities”, beginning the exploration of intelligent quantitative planning and governance decision-making in a big data environment. The integration of the multi-source big data and the construction of spatial models permits the simulation of complex urban systems, the better perception of urban signals, the monitoring of urban activities and the forecasting of urban futures.

2020 年，沉湖湿地开始进行智慧感知系统建设。在鸟类集中区域，设置了 32 个高清摄像头，对水质、土壤、鸟类、植物等湿地资源进行全要素、全周期的实时监测。设在管理局的监控室后台，可通过监控扫描识别鸟类种类和数量。大红鹳就是工作人员在 2021 年冬天通过监控发现的。实时的鸟类数据可有效反映保护效果。

沉湖“千里眼”监测设备　供图：缤纷自然
The “Clairvoyance” Monitoring Equipment on Chenhu Lake　Photo Credit: Colorful Nature

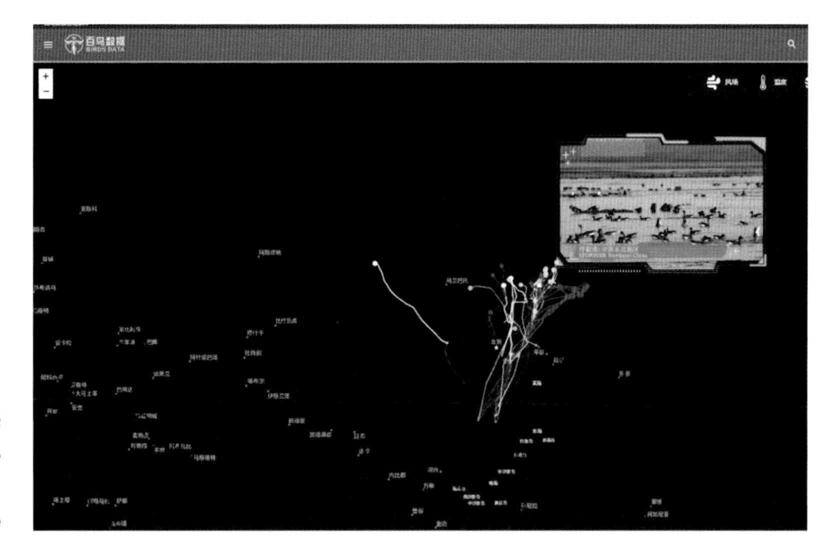

监测鸟类活动轨迹
供图：缤纷自然
Monitoring the Movement of Birds
Photo Credit: Colorful Nature

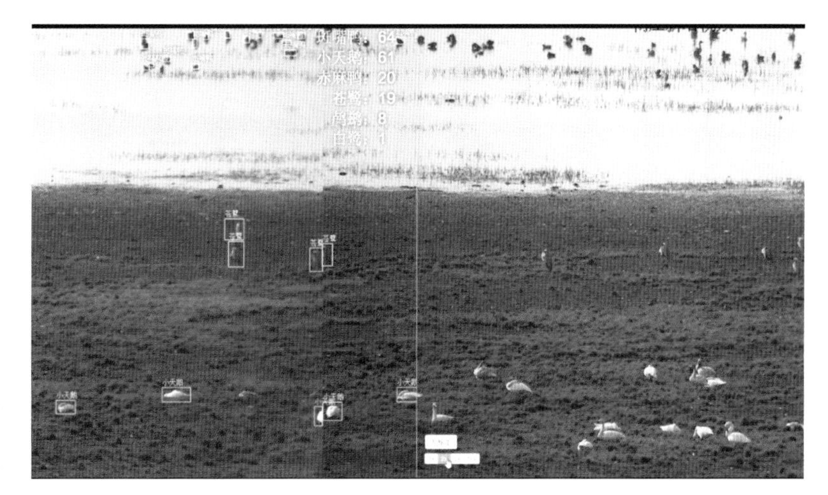

监测鸟类数量
供图：缤纷自然
Monitoring the Population of Birds
Photo Credit: Colorful Nature

通过声音识别物种
供图：缤纷自然
Identifying Bird Species by Sound
Photo Credit: Colorful Nature

In 2020, the construction of an intelligent sensor system in the Chenhu Lake Wetland was started. In the areas where the birds gather, 32 high-definition cameras have been set up for real-time, all-factor, full-cycle monitoring of wetland resources such as water quality, soil, birds, plants, etc. From the Management Bureau's monitoring room, the species and numbers of birds can be remotely scanned and identified. The greater flamingo was discovered by the staff via such monitoring during the winter of 2021. Real-time bird data provides a more effective reflection of the protection efforts.

保护区湖面还设置了多个浮漂，并配备有声音采集、水质、空气、水文等监测设备，实时反馈信息。

There are a number of buoys on the lake surface of the nature reserve which is also equipped with monitoring devices collecting real-time informatiom about sound collection, water quality, air quality, and hydrologic conditions, etc.

这套智慧湿地治理解决方案，通过在监测区域布设自运行、自供能、自传输，并且低成本、易维护的全要素感知设备，充分发挥已建成物联设备的潜力，实现了监测数据的全区覆盖、实时回传。

Through the installation of all-factor, self-operating, self-powered, self-transmitting, low-cost, easily-maintained sensing equipment in the monitoring area, this intelligent wetland management solution takes full advantage of the potential of IoT equipment to achieve full coverage and real-time collection of monitoring data.

同时，还将 AI 深度学习技术用于物种、人类活动的声纹图像识别，实现了对 100 多种物种和事件的自动精准采集与识别。自动收集的海量数据为公众提供优秀的数据资源，也为生态感知与监测应用提供强有力的数据支撑。

At the same time, AI deep learning technology is being applied to the recognition of the sound made by bird species and human activities, and helping automatically and accurately collect and recognize over 100 species and events. The huge amounts of data collected automatically provide an excellent public data resource, and strong data support for other ecological sensing and monitoring applications.

“千里眼”系统界面 供图：缤纷自然
The “Clairvoyance” System Interface Photo Credit: Colorful Nature

最终形成沉湖湿地数字孪生体，实现生态系统仿真、全要素监测和决策推演，协助管理者精准感知、智能响应和科学治理。利用更为智能化、科学化、精细化的管理手段和工具，监测过程从以人工采集为主进入以自动化采集分析为主的阶段，将“人管”转化为“技管”，全面提升沉湖湿地的数字化管理水平。

Finally, a digital twin of the Chenhu Lake Wetland has been created, offering ecosystem simulation, all-factor monitoring and decision inference to assist administrators with accurate sensing, intelligent response and scientific management. Thanks to more intelligent, scientific and refined management methods and tools, the monitoring process has transitioned from reliance on manual data collection to automatic collection and analysis, with “human management” shifting to “technological management”, representing an overall improvement in the digital management of the Chenhu Lake Wetland.

智慧感知自然，运用大数据驱动业务和管理创新，沉湖国际重要湿地力争建成国际领先的一流智慧湿地，并推进林草生态治理体系和治理能力现代化。

Sensing nature with intelligence, and using big data to drive services and manage innovations, the Chenhu Lake Wetland of International Importance is striving to develop into a world-class intelligent wetland, and to promote the modernization of forest and grassland ecological governance systems and capacities.

后　记 Postscript

草木迷人眼，湖水绿如蓝。底蕴深厚的云梦大泽，是生生不息的万物福地。芳草碧连天，鱼戏莲叶间，鹳鹤云上舞，平沙落秋雁。历经沧海桑田的变迁，古云梦泽余脉沉湖，如一颗夺目明珠，在江汉平原的大地上再现光芒，在守护城市生态安全的同时，传承云梦生态智慧成为传播生态保护理念、丰富城市形象、彰显城市魅力、提高人民幸福感的一大引擎。

With luxuriant vegetation, clear water, deep-seated beauty and enchanting landscape, the ancient Yunmeng Wetland historically served as the happy habitat for various wildlife species. And through the vicissitudes of time, Chenhu Lake, a remnant of the Yunmeng Wetland, has reappeared on the Jianghan Plain, shining like a dazzling pearl. Inheriting the ecological wisdom of the Yunmeng Wetland while preserving Wuhan's ecological security, it has also become an engine for spreading awareness of ecological protection, enriching the city's image, highlighting the city's charms, and promoting the people's happiness.

斑嘴鸭游戏于沉湖　李梓固摄
Chinese Spot-Billed Ducks Swimming in Chenhu Lake　Photo by Li Zigu

“江、河、湖、城”交相辉映的世界名城武汉，“珍爱湿地，人与自然和谐共生”也已成为共识，时时处处“为子孙谋、为全局计”，无不彰显生态、社会、经济、文化效应。闹市的流光溢彩与湿地的天光水影遥相呼应，大街小巷与湖潭曲径气脉通联，万鸟齐飞奏出大都市的田园牧歌，武汉人诗意栖居的梦想灿然成真。

Wuhan is a world-famous city, where rivers, lakes and cityscape add radiance and beauty to each other, and the consensus of “cherishing wetlands and pursuing harmonious coexistence of humanity and nature” has been achieved. Focused on “seeking the well-being of future generations and taking the overall situation into consideration” at all times, the ecological, social, economic and cultural achievements are becoming apparent. The splendor of the bustling city echoes in the wetlands; streets and alleys are intertwined with the wetlands through the winding paths; thousands of birds fly in the sky, singing the pastorale of the metropolis, bringing to life the aspiration of the Wuhan people for a poetic existence.

城市，倾心尽力呵护湿地；湿地，吐故纳新润泽城市。

As the city protects the wetlands with all its might, the wetlands nurture the city through constant inspiration.

世界，穿过湿地的层层涟漪认识武汉；武汉，以湿地的飞羽芳华惊艳世界!

The world knows Wuhan through its wetlands, and through its wetlands, Wuhan dazzles the world.

（鄂）新登字 08 号

图书在版编目（CIP）数据

生态沉湖：候鸟福地 / 武汉市园林和林业局组编 . — 武汉：武汉出版社，2022.11

ISBN 978-7-5582-5477-2

Ⅰ . ①生… Ⅱ . ①武… Ⅲ . ①沼泽化地 - 介绍 - 武汉 Ⅳ . ① P942.631.78

中国版本图书馆 CIP 数据核字（2022）第 171736 号

组　　编：武汉市园林和林业局

责任编辑：刘从康　明廷雄　　审　　校：游长松　王　熙

封面设计：金　蕊　　　　　　版式设计：夏玉洁

督　　印：方　雷　代　湧

出　　版：武汉出版社

社　　址：武汉市江岸区兴业路 136 号　　　邮　　编：430014

电　　话：（027）85606403　　85600625

http://www.whcbs.com　　E-mail:whcbszbs@163.com

印　　刷：武汉精一佳印刷有限公司　　　经　　销：新华书店

开　　本：787 mm×1092 mm　　1/16

印　　张：7.5　　　字　　数：188 千字

版　　次：2022 年 11 月第 1 版　　2022 年 11 月第 1 次印刷

定　　价：68.00 元

关注阅读武汉
共享武汉阅读